Making Mathematics Meaningful

For Students in Primary Grades

Werner W. Liedtke

Order this book online at www.trafford.com
or email orders@trafford.com

Most Trafford titles are also available at major online book retailers.

Cover Design and Designed by Karen H. Henderson

Printed in Victoria, BC, Canada.

ISBN: 978-1-4269-2344-9

*Our mission is to efficiently provide the world's finest, most comprehensive
book publishing service, enabling every author to experience success.
To find out how to publish your book, your way, and have it available
worldwide, visit us online at www.trafford.com*

www.trafford.com

North America & international
toll-free: 1 888 232 4444 (USA & Canada)
phone: 250 383 6864 ◆ fax: 812 355 4082

Table of Contents

To Linda.

Acknowledgements

A sincere thank you to Patty Kallio and to Judith Sales who read the manuscript and made many invaluable suggestions.

Since most of the ideas for activities and problems in the book came from experiences with students in many different classrooms a thank you is extended to the teachers who made these visits possible.

Over the years many students were interviewed.
Quite a few of these students agreed to be video taped.
The willingness of these students to talk about mathematics was not only appreciated, but very enjoyable as well.
A thank you goes to these students.

A special thank you also to Dorothy for her patience, encouragement and suggestions.

Chapter 1 – Teaching to Make Mathematics Meaningful

To be a Teacher – Selective Challenges

The content and suggestions in this book are based on an assumption voiced by a mathematics educator quite a few years ago,

> There is not now, never has been, and is hoped never will be a genuine substitute for a good teacher who knows how and what children need to learn and when they need to learn it. [1]

The term 'good' in the quote is rather subjective and as far as one of the goals of this book is concerned lacks specificity. The descriptors knowledgeable and skilful could be substituted but they require elaboration or need to be illustrated with examples. Different types of knowledge and a wide range of skills related to teaching, learning and assessment are required to make mathematics meaningful to young learners.

The delivery of the critical components that young students must encounter in a mathematics program and having these students reach the goals presented in the mathematics curriculum[2] requires having knowledge about:

- child development;
- general teaching strategies;
- specific strategies related to teaching about mathematics;
- a repertoire of assessment techniques;
- writing meaningful reports; and
- the mathematics content that is to be taught.

It has been suggested by some, 'that anyone who is qualified to teach one subject is therefore qualified to teach any subject.' [3] The author of the reference concludes that such an assumption is analogous to suggesting that a dermatologist and an orthopedic surgeon ought to be freely interchangeable. This seems absurd. Teaching about mathematics at any level requires specialized knowledge.

Is knowledge of the content alone, in this case mathematics, sufficient? The same author concludes that anyone who responds in a positive way to this question would agree to have someone with an anatomy degree practice medicine and declares, Education is at least as difficult as medicine, and is more important to society (if not to the individual.) Making mathematics meaningful, which implies fostering the development of conceptual understanding, requires specialized strategies and skills.

Making mathematics meaningful requires special teaching strategies, effective activity settings, meaningful practice, methods of accommodating all types of students' responses, different types of assessment techniques and skills for translating assessment results into meaningful reports. These requirements illustrate the key role played by a teacher. It will become aparent that this role cannot be replaced by any printed materials or by components of technology.

Everyone knows or seems to know something about teaching. The word 'teaching' is used in many everyday settings. During conversations with people about teaching it becomes evident that many people think they know a lot about teaching. They have, after all, been through the school system.

During conversations with representatives from different professions, be they pharmacists, dentists or lawyers, it becomes clear that each profession has areas of very specialized knowledge. If, by chance, an educator gets a turn to have a say about teaching, learning and assessment, everyone listening will want to make comments about education. Not only do they want to identify what they think is wrong with the education system and with teaching, they have definite ideas about what needs to be or should be done. The suggestions they make are likely to take them down memory lane to the good old days. It is not unusual to listen to a conclusion like, '*It worked for me. I am successful therefore it must have been good or effective.*' Responses to requests for elaboration many times include references to rote procedural skills that were learned and to results of memorization.

There are routines that are part of teaching which could easily be carried out by almost anyone who is not afraid of standing in front of a group of students. Routine tasks that are related to the management of a group include such things as taking attendance; establishing and reinforcing rules of conduct; asking simple and clear questions and eliciting responses; assigning tasks; and marking assignments. These tasks do not require any specialized knowledge. Teaching involves much more than dealing with routine tasks.

There exist ways of presenting mathematics content that may not require the specialized knowledge that has been suggested. In a teacher-centered or teacher-owned setting the knowledge comes from the teacher and/or from a text. The thinking is done for the students. Examples and rules are presented and these are followed by extensive practice. Teaching to make mathematics meaningful is much more than being a textbook wired to sound; it requires a role as a *coach of thinking*.

The *critical components* and goals of the mathematics curriculum go beyond memorizing facts and procedures. Having young learners reach these goals requires special strategies that teachers have at their disposal and make part of their teaching – learning – assessment settings. The strategies include:
- Fostering *self-confidence* in students.
- Fostering ability to *take cognitive risks*.
- Getting students to *think*, to think *flexibly*, and to *think about their thinking*.
- *Advancing thinking*.
- Fostering perseverance.
- Arousing *curiosity* and *use of imagination*.
- Allowing for and accommodating *spontaneity*.

Teaching to Make Mathematics Meaningful – Selective Challenges

It is surprising how many people there are who have a definite opinion, not only about how mathematics should be taught, but also about what is wrong with how it is presently taught. One of the first things old-timers will say is that, *'Today's students do not know their multiplication facts as well as we do. When we went to school ....'* Some will take great pride in the fact that they can recite the multiplication tables up to 16 or even 19. Teachers who as part of their teaching *'drilled these facts into us'* may be spoken of with some degree of fondness. It is not surprising that many will place the blame for this lack of knowledge in today's students on the calculator.

All parents have some ideas about teaching aspects of mathematics to their young children. Books about numbers and counting can be found in most homes and these topics become part of conversations, recitals and songs. The examples that follow are not intended to poke fun at anyone. The sole purpose is to indicate that serious misunderstandings about aspects of mathematics exist and that the mathematical ideas that are involved are much more contrived and complex than people think. The point is that the recital of number names, even if this is done in the correct order, has nothing to do with understanding number and understanding counting.

Many parents use the expression, *'teaching numbers'* or *'teaching to count'* as they explain what they are doing with their children. After sharing that his five year old son could count to thirty, one proud father declared, *'He knows his numbers, now he is learning his letters.'* Many parents equate the ability to recite number names in order with an ability to understand numbers. Such a conclusion is analogous to stating that the ability to recite the alphabet means that a child knows how to read and understand what is being read.

Three conversations with mothers of children who were almost five years old further illustrate attempts to teach. One mother explained that she was teaching her son how to add by having him practice with equations obtained from the internet. The second mother explained how she was teaching her daughter about subtraction by counting how many more bites are to be taken from the plate at supper time. The third mother was with her daughter in a driveway as they were entering numerals into a hopscotch they had drawn. According to the mother, she was *'teaching numbers.'*

These parents were serious about the statements they made and the goals they had in mind. It could be that some of these ideas had their origin in one of the many references that deal with number and counting that are available to parents in stores. These booklets tend to focus on low level cognitive tasks.

Teachers of young children have an important role to play. They need to convince many parents of young children that making the learning about numbers, counting, operations and other aspects of mathematics meaningful involves levels of high order thinking. Sense making is a key part of making mathematics meaningful.

Teaching about mathematics requires strategies that accommodate the *'critical components that students must encounter in a mathematics program in order to achieve the goals of mathematics learning'* and the *'main goals of mathematics education.'* [2] The critical components include ideas related to: *communicating*; *connecting*; *mental mathematics* and *estimation*; learning new mathematics

(cont'd next page ...)

through problem solving, *mathematical reasoning*, use technology as a tool, and the ability to *visualize*. The main goals include the preparation of students to: use mathematics *confidently* to solve problems, *communicate* and *reason mathematically*, appreciate and value mathematics and to *make connections*.

As discussions and activities are planned and while interacting with students strategies are required that specifically relate to the accommodation of the *critical components* and to reaching the goals for students. A list of sample strategies, in no particular order, can include:

- Keeping in mind the characteristics of good or effective problem solvers and accommodating these whenever possible.
- Developing ability to solve problems *via* or *through* problem solving.
- Using specific outcomes for planning lessons and activities.
- Planning for a balance of invention, demonstrations that include examples, non-examples, and appropriate practice.
- Enabling students to *connect* what they learn to previous learning, to ongoing learning and to their experiences.
- Developing key criteria for *mathematical reasoning*.
- Fostering the development of *visualization*.
- Developing the important aspects of *numeracy: number sense, spatial sense, measurement sense, statistical sense* and *sense of relationships*.
- Enabling students to talk and write about the mathematics they learn in their own words.
- Enabling students to connect mathematical terminology to familiar language.
- Enabling students to acquire estimation strategies and strategies for *mental mathematics*.
- Having students develop *personal strategies* for computational procedures.
- Using technology to foster important aspects of *numeracy*.
- Taking advantage of the power of high order thinking questions and open-ended questions and tasks.
- Using high order thinking questions and accommodating all types of responses by students during the orchestration of discussions.
- Recognizing the importance and need for specific goals, and specific and correct language.
- Providing appropriate practice settings that contribute to fostering important aspects of *numeracy*.
- Maximizing participation while orchestrating discussions.
- Using assessment questions and tasks that are appropriate and fair.
- Preparing specific reports for parents.
- Using the teaching about mathematics settings to contribute to language development, reading comprehension and the development of evaluative skills.
- Being aware of possible issues related to equity and multiculturalism.

Making mathematics meaningful for young learners involves many challenges. The examples that are included in the lists of strategies illustrate some of the challenges that a teacher faces and the members of these lists reiterate the importance of the role of a teacher.

The Purpose of the Book

The main purpose of the book is to make practical suggestions that can contribute to making mathematics meaningful for young learners and to do this with a minimum of theoretical discussion. The suggestions include:

- Types of questions that can be asked.
- Ideas for orchestrating discussions.
- Ways of accommodating different responses by students.
- Types of activities and problems for key ideas, procedures and skills.
- Examples of appropriate practice.
- Methods of assessing how meaningful the mathematics that has been learned is for young students.
- Ways of reporting assessment results for key aspects of mathematics learning.

The goal of making mathematics meaningful to students cannot be reached without the development of *number sense*, the key foundation of *numeracy*. Showing how this development of *number sense* can be fostered and illustrating how it is an essential requisite for key ideas, procedures and skills is an important part of the framework of the book.

Making mathematics meaningful for young learners requires special strategies. Throughout the book strategies are illustrated that can contribute to:

- Fostering the ability to *visualize*.
- Fostering problem solving ability in *through problem solving* settings.
- Fostering *mathematical reasoning* which includes *thinking about thinking* and *flexible thinking*.
- Fostering ability to make *connections*.
- Fostering *confidence*, *willingness to take risks* and *use of imagination*.

Although the focus of the content of the book is on attempting to make mathematics meaningful to young learners, the strategies and activities that are described and illustrated can contribute to students' *language development*, *reading comprehension* and to the development of *evaluative skills*.

For Reflection

Assume that you agree with the quote by Polyani[4], *'The teaching of mathematics is not only incredibly important, but also one of the most difficult topics to teach.'* What major points would you include in a presentation to illustrate the importance and the possible difficulties of teaching mathematics?

In 1989 Willoughby[2] wrote, *'Our world is becoming more mathematical. We are constantly surrounded by mathematical situations and are regularly required to make mathematical decisions. Muddling through mathematics without the appropriate attitudes and abilities in mathematics has become and will continue to become, more difficult – for both individual and society.'*

   a) What are possible examples of appropriate attitudes and abilities?
   b) What possible implications does this observation have for how mathematics is taught?
   c) What possible implications does this observation have for how mathematics learning is assessed?

How would you respond to someone who states that issues related to equity and multiculturalism are not related to aspects of mathematics teaching and learning?

Newspapers put their support behind, *'Raise a Reader'*. Consider possible reasons for changing this goal to, *'Raise a Numerate Reader who is able to Think'*.

Chapter 2 – The Framework and Assumptions

This book is for anyone who is interested in helping young students in the primary grades in their early journey of making sense of the many fascinating aspects of mathematics. The purpose is to share information that can be of assistance not only with reaching major goals of the mathematics curriculum but also valuable outcomes for other areas of learning.

Aspects of Meaningful Mathematics Learning – Main Assumptions

Conceptual Knowledge

Conceptual understanding facilitates transfer and therefore is the key to success for future mathematics learning. It is a pre-requisite for taking risks in problem solving settings and inventing personal strategies for computational procedures.

Conceptual understanding enables young students to look at dramatizations, simulations, or sketches that depict action and recognize the numbers, the action and the order of the numbers. This recognition enables students to use numerals and the appropriate symbol to record a matching summary or equation. For example, a picture showing two planes in line on a runway and one plane taking off is summarized by recording **3 – 1.** If *'taking off'* or *'flying away'* is used as part of a matching word problem, this is indicative of understanding the mathematical term *minus*.

Conversely, given a summary or an equation, *conceptual understanding* enables students to make up a meaningful word problem from their experience, simulate the action with objects and to prepare a sketch that shows the action. For example, for **3 x 4** the responses by a student of:
- reading this summary as *'three groups of four'*;
- showing three sets of four objects;
- creating the word problem, *'I am thinking of the wheels on three cars'*

are indicators of the ability to *visualize* or create visual images of the numbers, the order and the operation, and ability to connect to experience. V*isualization* and *connecting* are important indicators of *conceptual understanding*.

The ability to *connect* new learning to previous learning is another indicator of *conceptual understanding*. For example, a student's explanation that the answer for **7 – 5 = ■** can be found by thinking of, *'how many fingers I would have to add to five fingers to show seven fingers'* provides indicators of the ability to *visualize* numbers and to *connect* subtraction to addition.

(cont'd next page ...)

The goal to move from *'speedy, and too often meaningless mastery of paper and pencil computational skills to the development of numerical power'* [1] requires the development of *conceptual understanding* as a key component of mathematics instruction. Research results indicate *'that instruction can emphasize conceptual understanding without sacrificing skill proficiency.'* [2]

An emphasis on the development of *conceptual understanding* means that during the planning of lessons consideration be given to the development of students' abilities to *visualize*, *connect* and *communicate*. The results of such an emphasis will facilitate future learning of mathematics.

Problem Solving

Teaching *via* or *through* Problem Solving
In *via* or *through* problem solving settings students are faced with tasks that are new to them. Questions or requests of the following types may be presented in such settings:
- *Show how you could* use your fingers in two different ways, other than counting, to find the answer for **7 + 7 = ■**.
- *How could you try* to find out how fast the water in a creek is flowing?
- *Use what you know to try* to find the answer for ...
- *How would you ...?*

Attempts to satisfy these types of requests require *self-confidence* to take cognitive risks; *conceptual understanding* of what has been learned; and ability to make sense of numbers.

The following example illustrates a *via* or *through* problem solving setting in a grade two classroom. The general goal is to have students develop a *personal strategy* for the addition of two-digit numerals.

The following assumptions are true. The students:
- must have *conceptual understanding* of the operation of addition as explained in the previous section.
- must have developed *mental mathematics strategies* for the basic addition facts to either show that answers they state are correct or to re-invent answers that may have been forgotten.
- are able to *visualize* numbers to ninety nine. When students look at two-digit numerals they can explain how these numerals can be represented with the fewest number of base-ten blocks or with the fewest number of children showing the numbers with their fingers.
- are able to *think flexibly* about numbers. The students know that a number can be shown and named in different ways.

The following problem is presented to the students:

> *Last month 17 books were ordered.*
> *This month 25 books were ordered.*
> *How many books were ordered in these two months?*
> *Try to think of at least two different ways to figure out the answer.*
> *Use ways other than counting.*
> *Be ready to explain your thinking and to show it with fingers of students.*

(cont'd next page ...)

When students are requested to attempt this task with a partner, *communication* is part of the setting. The explanations of the strategies that are to be prepared and shared with the members of the class contribute to language development. The open-ended request accommodates individual differences and allows for further exploration. Decisions will have to be made when different strategies are reported and similarities and differences are discussed.

For the majority of students the *personal strategies* that are invented in these settings differ from the standard procedure or standard algorithm that has been taught in classrooms. Most students prefer to begin their procedure with groups of ten.

An assignment of trying two similar tasks in two different ways and being ready to state which of the two is preferred and why, fosters flexible thinking and involves further decision making.

This example illustrates one setting for having students who are *self-confident*, have *conceptual understanding* and a *sense of number* develop *personal* computational procedures for finding an answer The importance of the role of the teacher is reiterated since discussions need to be orchestrated and different responses need to be accommodated.

As these types of settings are planned and orchestrated the accommodation of characteristics of successful problem solvers (see chapter 3) needs to be kept in mind.

Mathematical Reasoning

Mathematical thinking is in many ways like everyday thinking.[3] One important criterion of this thinking is that *everything should make sense*. This sense making, and the development of *mathematical thinking* in general, can be fostered by using well-phrased and well-placed high order thinking questions during lessons.

Observations and research data indicate that the majority of questions asked during mathematics lessons tend to be low order thinking questions. Certain settings make it easy, convenient and perhaps tempting to ask many questions of this type. It is also true that the inclusion of high order thinking questions does require extra time during the planning stages.

Other criteria of *mathematical thinking*[4] that are part of the framework of the book include:

- **Ability to get oneself unstuck.** Students will acquire *mental mathematics strategies* that will enable them to re-invent ways to reach answers that have been forgotten.

- **Ability to correct errors in answers, in the use of materials and in thinking.** During demonstrations, which are one of three parts of an effective balanced mathematics program, students can be given opportunities to react to these types of errors and speculate about possible reasons for these mistakes. This speculation can contribute to fostering students' *conceptual understanding*.

(cont'd next page ...)

- *Performing calculations with a minimum of counting and rote pencil and paper computations.* The assumption is made that counting is not a mental mathematics strategy. Rote learning is an oxymoron.[5] Rote pencil and paper calculations are not part of the goals of mathematics teaching and learning.

- *Willingness to try another strategy instead of giving up.* It is better to solve any problem in more than one way than practicing to solve many problems in the same way. Many of the questions that are suggested in the book invite students to go beyond one response. Strategies are presented that encourage *flexible thinking*, even the use of *imagination*.

- *Ability to extend a problem situation by posing additional questions.* The best that teachers can hope for is to encourage this ability whenever signs of it appear. This encouragement is kept in mind, pointed out and sample questions that might contribute to reaching this goal are suggested when the opportunity arises.

Number Sense

Number sense is the key foundation for successful mathematics learning. It is essential for developing **numerical power**.

Sense of number enables students:

- **to make estimates about number.** Students will learn to use *referents* and *benchmarks* as part of their estimation strategies and as part of making statements about numbers that involve the term *about*. The rote 'rounding procedure' is not part of the ability to make estimates.

- **to develop *mental mathematics strategies* for the *basic facts*.**
  These strategies require being able to *visualize* and to *think flexibly* about numbers, key aspects of *number sense*.

- **to develop *personal strategies* for calculation procedures.**
  As the example for the *via* problem solving setting illustrated, *number sense* and *conceptual understanding* are pre-requisites for being able to develop these *personal strategies*.

- **to make predictions about answers.** Students need to learn when predictions suffice and calculations are not required. The ability to *visualize* numbers and having *estimation strategies* at their disposal will enable students to trust their predictions.

- **to be able to assess the reasonableness of answers for calculations.**
  Without *number sense* and without being able to estimate, reasonableness of answers cannot be assessed.

Without *number sense* mathematics learning becomes rote and meaningless. Special teaching – learning settings; special use of materials; high order thinking questions; and well planned strategies are required to foster the *visualization of number* and the *flexible thinking* about numbers that are key aspects of the development of *number sense*. Without this *number sense* the goals set out for students in the mathematics curriculum will not be reached.

Spatial Sense

Visual imagery and *visual thinking* are important components of s*patial sense*, which includes *measurement sense*. Since the solution of many mathematical problems requires aspects of *imagery* and *visualization*, *spatial sense* is an essential aspect of problem solving ability.

Spatial sense is essential in many tasks. These include such things as:
- printing and writing letters and numerals,
- reading and constructing tables for information,
- following and writing instructions,
- sketching and interpreting diagrams,
- reading and drawing maps,
- visualizing objects and locations that are described verbally,
- visualizing pictures in three dimensions.

Spatial sense can be developed and improved.[6] As is the case for *number sense*, classroom settings which foster the development of *spatial sense* require skilful planning. Well-phrased and well-placed high order thinking questions that contribute to the development of *visualization* are required.

Since *spatial sense* is not only an important aspect of *numeracy*, but is part of many other areas of learning an argument can be made for making geometry the first topic students encounter at the beginning of the school year and then making it part of teaching whenever possible.

Statistical Sense

Students learn how to interpret data plots and graphs. Discussions about the role of appropriate titles and the identification of major ideas that are displayed are part of this interpretation. Students also learn how to classify statements about displayed data as:
- *true*;
- *not true* or *false*; and
- *could be true* or *false* – but the display does not show.

Aspects of language development, comprehension and evaluative skills are accommodated when students create and record their own statements about data displays that are:
- *true* or *certain*;
- *false*;
- *likely* or *probable*; and
- *not likely*.

Data plots and graphs can be used by students as a problem solving strategy. The organized data of the information that students collect in response to questions or concerns can lead students to the formulation of answers and to the drawing of conclusions.

Sense of Relationships

Students will have opportunities to think about patterns not only as part of readiness for counting with understanding or rational counting, but in problem solving settings that allow for flexible thinking and the use of imagination. Students are asked to make generalizations and test these when opportunities arise. New learning is related to every day actions or events from the students' own experiences.

Calculators

Calculators can be used in ways that require thinking or flexible thinking. Tasks can be designed that foster the development of important aspects of *number sense*.

Research shows that calculator use does not harm students' achievement and that students who use calculators possess better attitudes and self-confidence toward mathematics than non-users.[7]

Specific Goals and Outcomes

Specific goals are needed not only to provide a focus for any activities and settings, but also to make assessment possible that is non-subjective and results in reports that are free of general language, or 'edu-speak.'

The strategies for the presentations, the types of questioning, and the suggestions for accommodating possible responses by students that are described in the book are intended to illustrate how the specific outcomes that are stated might be reached. However, the intent of these strategies and settings is also to contribute to fostering a high level of *confidence*; *risk-taking*; *perseverance*; *curiosity*; even a *tolerance for ambiguity*; and *imagination*.

Students and Mathematics Learning

Students' mathematics is perceived as an emergent event. The mathematics students learn is a result of engaging in settings that involve physical, verbal, and mental activity as individuals and as members of a group. Mathematical learning is a process of continuous growth. Teachers can nurture this growth. *Mathematical thinking* and *mathematical understanding* involves much more than a connection of linear ideas and skills. New learning is related to previous learning, to ongoing learning as well as to events from every day experiences. *Mathematical understanding* is always emerging and changing. Since that is the case, any assumptions about students' levels of understanding of mathematics at any one stage are not easily made, nor are they easy to predict.

Mathematical Language

During the preparation of the content of this book a conscious attempt was made to avoid language that may be difficult to interpret, that is open to possible misinterpretations, or may even be impossible to interpret. An attempt was made to use language that is mathematically correct. This last statement may seem unusual, but mathematical language is used incorrectly by many people, even in some references.

The ability to recite mathematical terms and phrases used by an adult without understanding their meanings is not an indicator of *mathematical thinking* or *mathematical understanding*. The acquisition of appropriate and correct language is important but it is not the aim when learning about something new. Whenever possible, discussions with students about mathematical ideas and procedures begin with natural or everyday language. In most instances, it will take time for students to learn and use actual mathematical terminology correctly as part of their conversations and writing.

The mathematical language used in conversations with students has to be appropriate and correct. This requires caution when using terms that have more than one meaning or use. A list of possible examples includes words commonly found in everyday conversation. These include: *number* and *amount*; *number* and *number name* (numeral); *pattern* and *design*; *guess* and *estimate*; *figure* and *shape*; and assigning names of two-dimensional figures to three-dimensional figures – for example, calling a *cube* a *square*. As part of everyday usage the term fraction can have a different meaning than the one used when the topic is introduced to students.

It is also beneficial that those who are working with the young students have correct and consistent meanings for mathematical terms. For example, the meanings for the following terms *basic fact*; *game*; *problem solving*; *triangle*; *rectangle*; and *circle* could differ among participants while taking part in a discussion or even for authors who write about mathematics teaching and learning. Students also need to learn to use the correct language when comparisons are made that involve numbers or amounts. Terms like *fewer, more, greater than, less than, greatest, least*, or *most* are often used incorrectly by many people.

Organization of Topics

Whenever possible, an *open-ended approach* is used and suggested. Many examples of open-ended tasks and problems are included. Students can find such tasks mathematically inviting and challenging.

The strategies and activity settings are intended to provide important aspects of mathematics and *mathematical thinking* as prompts for students to wonder about, to question their current thinking in relation to the idea or ideas they have of these aspects, and to try and get them to think more deeply about these ideas.

Mathematical activity is not an isolated event. Aspects of mathematics are part of the everyday lives of students. Despite a focus on students' learning of mathematics, many of the *basic thinking strategies* that are dealt with in the book are not exclusive to the domain of mathematics. Some activities will indicate how these *thinking strategies* connect with other important areas of learning. This connection illustrates that the learning about mathematics can contribute to fostering the development of language, reading readiness, reading comprehension, and evaluative skills that are part of other subject areas.

Content and Parts

- The important *goals* for each topic are identified.

- *General learning goals* for problem solving are stated.

- Each topic includes suggestions for the orchestration of discussions and questioning that deal with important ideas and skills related to the topic. These strategies can be transferred to settings other than those used in the examples.

- Questioning strategies and suggestions for conversations are included. Whenever possible, *open-ended questions* are used and teaching *via problem solving* is illustrated.

- Suggestions are made for accommodating all types of students' responses.

- Since *documentation* plays an important role as part of fostering *thinking, flexible thinking, thinking about thinking* and attempts to *advance thinking,* suggestions for its use and role will be included whenever possible.

- Ideas for collecting assessment data are suggested that can be part of ongoing *documentation.* Any comments that are recorded about a student can be used to create a collection of 'snapshots' of *mathematical understanding.* Over time, these 'snapshots' can serve as indicators of growth that has taken place.

- Suggestions are made for reporting results of assessment.

- A few ideas for personal reflection or for discussion are included.

Fostering the development of *mathematical thinking* and *mathematical understanding* is a challenging journey, a journey that can be interesting as well as enjoyable - for both parties, the students as well as the persons who are in charge.

For Reflection

From time to time people without the appropriate credentials are able to fake being somebody they are not and they are quite successful at it since it may take some time until they are found out. Assume someone fakes to be a teacher of mathematics. For what types of settings might it be very difficult to determine whether the person is a fake? Are there any types of settings that have a high probability of identifying the person as a fake?

What could or should be said to a parent who claims that young students can learn mathematics from a computer?

A few students will make a statement about one very special teacher who taught them and who was able to make them understand mathematics? What might be special about such a teacher?

A newspaper report included a report about a student who was told by her favourite professor that she was 'too smart for teaching and she should consider doing something else.' What questions would you ask of this professor? What comment would you make?

What would you say to someone who states, 'In school I always liked mathematics because I knew it was always a matter of right or wrong.'

Chapter 3 – Teaching via Problem Solving

Problem Solving

Problem solving can be thought of as what happens and what is experienced when a situation is encountered where one does not know what to do. *Problem solving* is connecting and applying one's knowledge to new or novel situations, or in new ways to familiar settings. *Problem solving* is re-inventing or re-constructing something that has been forgotten.

Problem solving needs to be part of ongoing teaching and learning. Trying to deal with *problem solving* as a separate entity; trying to tell students how to solve a problem or giving them a plan to solve a problem; or providing key words and hints will not help students become *successful problem solvers*.

Sense making, or having everything *make sense,* is a requisite *for problem solving.* Without a *sense of number* and without *spatial sense* it is unlikely that the goal of *fostering the ability to solve problems can be reached. This sense making* is a key component of the activities illustrated in this book. However, the activities alone do not suffice. The *type of questions*, the *accommodation of all answers,* and the *conversations* that are suggested and are part of the activities are of prime importance. The *orchestration* of the activities provides the key component for teaching *via* or *through problem solving.*

The accommodation of all types of responses from students in a teaching *via problem solving* setting can not only promote *thinking* and *flexible thinking*, but *thinking* can be advanced as well.

General Goals for Problem Solving

Since *problem solving* is not a separate entity that is developed in an isolated setting, it does not make sense to suggest specific learning outcomes or goals. However, there are some key characteristics that students who are successful problem solvers possess.[1][2] These characteristics can be used to identify general goals for fostering the ability to solve problems and they provide guidance for the design and make-up of the activities and questions that are included as part of the text.

Successful problem solvers are able to note likenesses and differences.
The mental activities of *sorting* and *classifying* are not just an important part of *readiness for number* settings, but an aspect of the development of *number sense, spatial sense, measurement sense, and statistical sense* as well as other areas of *sense making* in mathematics.

(cont'd next page ...)

Successful problem solvers are able to visualize.
> The ability to *visualize* is one of the key components of *conceptual understanding* and making *sense of numbers*. Activities that get students to recognize number *without counting* (subetizing), thinking about numbers when number names are heard or numerals are seen (whole numbers; fractions; decimals), and thinking of different ways of showing the same number are all part of an attempt to foster *visualization*.

> *Spatial sense* activities include looking at blocks from different viewpoints, matching blocks with pictures of blocks and trying to describe the parts that can be seen and trying to make predictions about the parts that cannot be seen. These tasks along with having students consider more than one possible response for one stimulus are all examples of attempts to develop the ability to *visualize*.

> *Measurement sense* activities include looking at attributes of objects and learning that our eyes may deceive us.

> *Statistical sense* includes the inspection and interpretation of organized data.

Successful problem solvers are able to generalize on the basis of a few examples.
> *Generalizations* and testing these generalizations are essential aspects of mathematics. Students get an introduction to making generalizations and applying them when the simple properties for the *basic facts* are discussed, examined and tested.

Successful problem solvers understand mathematical terms and ideas.
> One way to foster the ability to use and understand *mathematical terminology* is by beginning with familiar language and then using it in conjunction with the mathematical terminology. This strategy is illustrated when students are introduced to new ideas, new procedures and new ways of looking at objects and attributes of objects. Students are not asked to memorize terms for the sake of doing so.

Successful problem solvers are able to connect.
> The ability to *connect* what is being learned to settings outside the classroom, to previous learning and to ongoing learning can be considered a problem solving strategy and a key indicator of understanding what is being learned. Whenever possible, the materials that are suggested for activities should be familiar to students and natural language should be used to introduce mathematical ideas and terminology.

Successful problem solvers have the ability to e*stimate*.
> The difference between guessing and *estimating* is explained while numbers are examined and discussed. Students learn to use *referents* and *benchmarks* as part of their estimation strategies.

(cont'd next page ...)

Successful problem solvers switch methods readily.

The reason why it is better to solve a problem in different ways rather than practicing the same solution procedure is illustrated throughout the book. Since young students may not have many strategies at their disposal, it would perhaps be more appropriate if the latter part of this characteristic reads, *are willing and ready to switch strategies*. This ability can be fostered by accommodating all responses and ensuring that the students learn to realize that many times there exists more than one way of doing things and there may be different answers for a question or problem. Examples in this book illustrate strategies and types of questions that can be used to create settings which are favourable for reaching this goal.

Successful problem solvers have high self-esteem.

Having a high self esteem will result in students *taking cognitive risks*. This *risk taking* is essential for:
- being willing to switch methods;
- trying different approaches;
- asking questions;
- accepting that the responses of others can be as correct or as acceptable as one's own responses;
- being willing and able to talk about what is learned in one's own words.

It is too easy to employ strategies in the mathematics classroom that are not conducive to fostering self-confidence. Some of these can include strategies such as:
- marking or assessing answers as right or wrong;
- asking low order thinking questions that require preconceived correct one word answers;
- emphasizing speed;
- marking items that have not been attempted as wrong;
- expecting specific steps to be used for computational procedures and to solve problems.

Self esteem is fostered in a setting where all explanations are valued and discussed and where the emphasis is not on reciting or repeating ideas in a specified way. The types of questions and the suggestions for accommodating responses that are part of many of the activities and tasks were designed to encourage students to take risks and to foster self-esteem. Any evidence of risk taking by young students needs to be praised. Such an acknowledgement can lead to further risk taking and, in turn, build confidence and self-esteem.

It easy to see that a delicate orchestration of conversations, along with trying to accommodate as many of the characteristics of successful problem solvers as possible, makes it impossible for a computer or for printed text to reach the desirable goals related to *problem solving*. Teachers play the key role! The suggestions and ideas included in the book are intended to assist teachers in their attempts to fulfill this important role.

Problem Solving – 'We've Come a Long Way ...!'

At one time activities in classrooms focused on teaching about and for problem solving. As part of this approach students were requested to:
- read word problems,
- write equations for word problems,
- perform the calculations,
- write answer sentences.

A prescribed assessment procedure allocated five marks for these tasks, two for the correct equation, two for showing the correct calculations and one for the answer sentence. In many pupil texts, a page of word problems predictably dealt with the last skill or procedure that had been examined.

On many pages of word problems, the first example included the equation that was required. Many students 'solved these problems' without ever reading them. They did not have to. The thinking had been done for them. On prescribed tests students were asked to use the same steps. The marks on these types of tests determined students' marks for problem solving on report cards. A very simple procedure but not a meaningful result.

Two researchers[3] conducted a study that divided K to 6 students from several schools into three groups. The control group followed the standard textbook program and solved only the problems in the textbooks. The members in one experimental group were excused from all textbook word problems. The members of the third group were similarly excused, but whenever a new skill or procedure was introduced during the year they made up their own 'mathematics stories' in accordance with the rule that 'you have to be able to do the arithmetic yourself before you can use it in a story' (p. 3). The results were summarized as follows:

> At the end of the year, both experimental groups did far better on standardized test word problems than the control group of children, who had been practicing textbook word problems all year. The children who had made up all their own applications scored dramatically higher in a test of applications skills than the other groups (p.21).

It seems that the only way to explain these somewhat amazing results is that it must have been that the children in the control group had been part of a mathematics program that defined problem solving as was described in the first paragraph, as a fixed procedure. This type of setting does not develop the ability to solve problems. The results of the study do point out the value of having students write their own problems. In this setting students are given opportunities to reflect and to connect what they have learned to their own experiences.

There was a time when teachers taught students specific steps to solve problems, *'The first step is ...; the second step is ...;'* etc. It is difficult to believe that as part of assessment techniques some teachers required students to reproduce the exact steps they were shown in order to receive full marks for the items on a test. There exist case studies where students did get all of the answers correct, but failed tests because they used their own solution procedures.

(cont'd next page ...)

There was a period when the idea became popular that students learn how to solve problems by giving them many different problems to solve – labelled by some as the problem solving mania period. Publishers and Ministries produced references of problems and sometimes charts that listed problem solving strategies for students. These charts appeared on many classroom walls. Some of these charts actually identified a first step for students that stated: *Understand the Problem*. What possible meaning could such a statement or request have for young students? This request is completely meaningless for students who lack important aspects of *conceptual understanding*, *number sense* and ability to *visualize*.

The problem solving references resulted in numerous requests for workshops. Many of these sessions had something in common. Teachers were asked to try to solve the three colour problem; a problem about the seven bridges in Königsberg and, of course, the hand-shake problem. After attending several sessions, one teacher was overheard sharing the observation, '*What do I do after the handshake problem?*'

There existed references that assumed that students learn how to solve problems by providing them with key words to give them hints about solution strategies. One author asks the question, '*What's wrong with this procedure?*' and answers it with, '*Almost everything.*'[4] He challenges those who believe in this strategy to present the following problem to students in grades three and four:

> *Mary walked 11 metres north.*
> *She then turned and walked 7 metres west.*
> *Did she turn right or left?*

The author suggests that, '*If the most common answer is 4, you will know that the class members have mastered the key word procedure (since "left" always means subtract) but are not reading and thinking about the problem*' (p.39).

During one conference the same author was approached by a teacher who asked what he would recommend to teach problem solving. His response was, '*A teacher.*' The teacher who posed the question could and should be asked to explain the meaning of, '*to teach problem solving.*' It is amazing the different meanings that are attached to this comment.

During a NCTM conference presentation, President Price reported research on nonsense problems. He stated that students faced with a problem that does not make sense begin to solve it anyway because they have concluded that mathematics does not necessarily have to make sense to them. Students did try to find an answer for:

> *A shepherd has 125 sheep and 5 dogs.*
> *How old is the shepherd?*

Students are fortunate to find themselves in settings where teachers keep in mind the characteristics of successful problem solvers and attempt to foster the ability to solve problems *via* or *through* problem solving settings. These settings will pay dividends for students as they learn more about mathematics and solve problems in the future.

Assessment Suggestions

Specific suggestions for assessment require specific goals and specific learning outcomes. Since the goals that are identified for problem solving are general, the suggestions for collecting observational data are of a general nature. Nevertheless, these observations are very important.

CHARACTERISTICS OF SUCCESSFUL PROBLEM SOLVERS:

As students share ideas and make comments during conversations with the teacher or in co-operative settings with a partner or in a small group, these comments can yield indicators about:

- *Willingness to talk using familiar and mathematical language.*
- *Willingness to try things and to try different things.*
- *Thinking and thinking about thinking.*
- *Flexible thinking.*
- *Advanced thinking.*
- *Being curious.*
- *Being spontaneous.*
- *Being able to connect.*
- *Use of imagination.*
- *High self-esteem.*

Any observations that are made about these favourable characteristics need to be entered into students' portfolios in the form of anecdotal comments or as part of a checklist.

PROBLEM SOLVING STRATEGIES:

The discussions students have with one another or in small groups and the reports they prepare and deliver will reveal data about the strategies that were employed to solve problems. For example, students may utter such statements as,

'We drew a picture' or, *'We acted it out with blocks.'*

The only list of problem solving strategies that is of any value to students is a personal list of such strategies. Since that is the case, the strategies could be labelled, i.e., preparing a diagram; simulation with objects, and entered into the portfolios of those who reported using them.

It is more important for students to be *curious thinkers* who use their *imaginations* and *love to learn*, than to have memorized numerous isolated skills. It is hoped that students will encounter many teachers along the way who will care more about *willingness to take risks*, *eagerness to ask questions* and *high self-esteem* rather than about how much has been memorized and the speed of responding to stimuli.

Throughout the book suggestions are made for how the development of the favourable characteristics of problem solvers might be fostered and maintained. The open-ended questions and tasks that are part of many suggestions should provide opportunities for students to experiment with new and different problem solving strategies.

Reporting

Fostering the ability to solve problems is the major goal of all mathematics teaching and learning. The importance of this goal makes it essential that assessment data related to problem solving be collected and the results of such collections be summarized and shared with parents.

During conversations with groups of parents the majority initially agreed that they would be satisfied or even happy to receive a comment like *'is a good problem solver'* about a student. When they were asked to explain what they thought such a comment might mean and to try and state specific examples of *'good problem solver'* most of the parents changed their minds. Those who did provide examples for what they thought *'being a good problem solver'* means included the comments:

- *Gets the answers right.*
- *Knows what to do.*
- *Knows the steps to find an answer.*
- *Knows his stuff.*
- *Knows how to find answers.*

It is very easy and it may even be tempting to make an assessment comment like *'is a good problem solver'* but it is obvious from the comments made by these parents that this observation is much too general to be meaningful to anyone who reads it. The same conclusion is true for similar statements about other topics of mathematics learning. For example:

- *Has good number sense.*
- *Makes reasonable estimates.*
- *Shows computational fluency.*
- *Knows the basic facts.*
- *Has spatial sense.*

Listening to students as they share comments with a partner or in a group or deliver reports to classmates will yield indicators of some of the characteristics of successful problem solvers and about the problem solving strategies that were employed. A possible checklist of characteristics of good problem solvers could include:

- Willingness to talk about what has been learned and is being learned in familiar and mathematical language.
- Indicators of willingness to take cognitive risks.
- Evidence of thinking and thinking about thinking.
- Indicators of flexible thinking.
- Indicators of ability to connect.
- Incidents of high self-esteem or self-confidence.
- Signs of use of imagination.

Entries in a checklist of this type make parents aware of the characteristics that have been observed. Those who take care of the students can then receive statements in the form,

> *'... observed indicators of ...'*

If during the year more observations are added, a statement could be made about the observed growth that has taken place. If, on the other hand, there exists a lack of entries for some important characteristics, parents should be informed about what they might do to try and foster the development of these characteristics. During meetings with parents they could be informed about what they might do and what they should not do to foster development of the favourable characteristics of problem solvers.

Whenever possible, the personal strategies students use need to be recorded in that student's portfolio. These entries of a list of personal problem solving strategies for a student can be used as indicators of the ability to think flexibly. As more entries are added throughout the year, they can be used as indicators of growth that is taking place.

The entries in a portfolio make it possible to share statements about problem solving ability in the form of,

> *'... have observed ... problem solving strategies that were used ...'*

A final report at the end of the school year can make reference to the 'growth' that took place during the year by providing a count of the strategies that were observed.

For Reflection

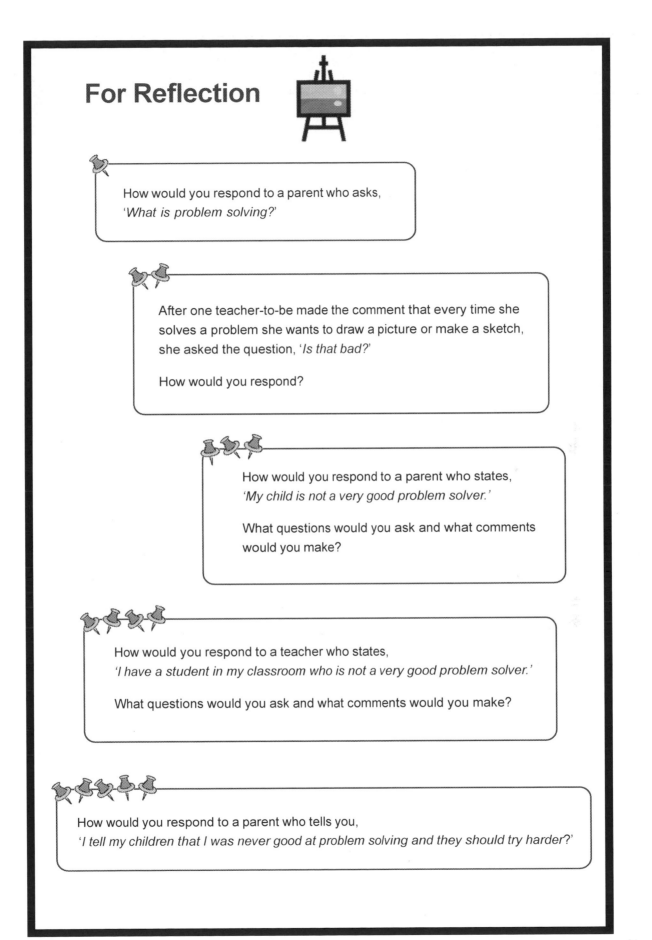

How would you respond to a parent who asks,
'What is problem solving?'

After one teacher-to-be made the comment that every time she solves a problem she wants to draw a picture or make a sketch, she asked the question, *'Is that bad?'*

How would you respond?

How would you respond to a parent who states,
'My child is not a very good problem solver.'

What questions would you ask and what comments would you make?

How would you respond to a teacher who states,
'I have a student in my classroom who is not a very good problem solver.'

What questions would you ask and what comments would you make?

How would you respond to a parent who tells you,
'I tell my children that I was never good at problem solving and they should try harder?'

Chapter 4 – Readiness for Understanding Number and Rational Counting

What is Number?

Think about driving through a rural area and seeing five horses grazing side by side. Next to one of the horses are five baby ducklings huddled together.

Young students would find it easy to look at these animals and talk about how the members of the two groups are different. Obvious differences include size, shape and colour. The arrangements of the animals in each group also differ.

Detecting and describing possible similarities for the animals in the two groups is not as easy for young students. They may talk about similarities like, *'they are animals'* and *'they have eyes'*, but recognizing that the numerousness or the number for the two groups is the same is another matter.

It is quite an achievement for young students to be able to say that, *'There are the same number in each bunch because there is a duckling for each horse'*; name the number, *'There are five'*; and then select or print the name for the number – **5**.

All sets of discrete objects whose members *match* or can be put into *one-to-one correspondence* have the same answer to, *How many?* The answer to this question is the *cardinal number* for these sets and the names that are assigned to these sets are labelled *numerals*.

Readiness for Number

The ability to categorize classes based on numerousness or number requires a high degree of *flexibility*. Students need to learn not to get distracted by the colour, shape or size of objects or by the way the objects in different sets that are compared are arranged. They need to learn to recognize that number can be used to establish classes that are the same.

Students also need to learn that it is possible to find 'the same number of' things or 'just as many' without having to count. Understanding the results and using matching or establishing one-to-one correspondence between objects of different sets is an important part of recognizing number as a common property and understanding number.

Sorting and Classifying

The major goal for sorting activities is to have students become very *flexible* as they examine and talk about how objects in their environment are the same and how they are different. The goal related to *flexibility* is to ensure that students are not distracted by the colour, shape or size of objects nor by the way discrete objects in different sets are arranged as comparisons are made and number is considered as a common characteristic.

The mental strategy of sorting and classifying is not just important for understanding number but is part of many other mathematical ideas as well as other areas of learning.

One typical strategy for sorting and classifying tasks consists of presenting four discrete objects or pictures and posing either of the questions:
>   *Which of these things is not the same?*
>   *Which of these is different?*

For example, students look at:
- four toy animals or pictures of these animals:
  cow; **dog**; **alligator**; **chicken**
- four numerals: [**4**]   [**5**]   [**6**]   [**10**]

The handling of students' responses during a discussion, on an activity sheet or as part of an assessment task can turn this into a 'closed' or 'heavy handed' setting if one specific choice is considered and marked as the correct answer.

In a 'closed' setting students have to try to guess what the person who designed the task had in mind and what the person was thinking. For example, students who did not select the alligator because it is a wild animal or the **5** because it is the name for an odd number could have their responses marked as being incorrect.

The Power of Open-ended Tasks and Appropriate Requests

A simple change of a question from
>   *Which one is different?* to, *Which one do _you_ think is different?*

with the emphasis on the _you_, has many desirable benefits and outcomes. Students learn to realize that whatever they think and report is accepted as correct. This, in turn, can build *confidence* and encourage students to *take risks*.

As part of an attempt to foster *flexible thinking* is concerned, it may be tempting to ask students, *Is there another one that is in some way different?* This type of low order thinking question has a great disadvantage. It invites a one word response and provides the opportunity to opt out rather than to think about another possible answer. If that is the case, the opportunity to continue the task and a conversation is gone.

Modifying the requests can encourage students to continue with the task on hand and re-examine the objects or things. For example:
>   *Try to think of another one that you think is in some way different.*
>   *Tell how it is different.*

Research shows that these types of open-ended questions and requests can foster *confidence* and *risk taking* in young students. [1][2]

(cont'd next page ...)

As far as *flexible thinking* is concerned, one goal for many settings and tasks would be that students can consider a wide range of possible attributes and eventually be willing to try to tell how every member of a collection might be different from the others. What might students who are very *flexible* say about each of:

- the four animals, **cow**, **dog**, **alligator**, **chicken** or,
- the four numerals: **[4] [5] [6] [10]** ?

Each new difference that students are able to describe for a member of a collection implies that new similarities for the remaining members are discovered.

Flexible thinking is fostered during many group discussions. As one student describes one difference that others have not considered, comments like the following are indicators of this desirable outcome:

> *'Oh yeah.'* or, *'I never thought of that.'*

If the intent is for students to identify a predetermined member of a set and they are unable to do so a change in strategy and questioning is required. As one object or picture is pointed to, students are told,

> *I think this one is different in some way.*
> *Try to guess what I am thinking of.*

If needed, appropriate hints can be provided, one at a time.

 Documentation can play an important role in attempting to foster *flexible thinking* and trying to *advance thinking*. If, for example, students working in pairs have sorted a given set of objects in several different ways, they are asked to make a list of what they were able to do or the list is prepared for them. This list can be presented to the same students at a future date along with the requests,

> *This is what you discovered last time.*
> *Try to think of some other ways to make groups that are in some way the same.*

The new sorting strategies are added to the list.

 ### *Types of Activities and Problems*

A classroom setting provides numerous opportunities for sorting and classifying tasks. These can include:

- Pictures students have drawn.
- Creatures students have constructed from modelling clay.
- Name tags.
- Title pages of books.
- Stories students tell.
- Pictures of animals.
- Buildings made with blocks.
- Designs made from construction paper.
- Word problems written for summaries or equations.
- Strategies they describe for solving problems.
- Different shapes or bridges they can make with their bodies.

Whenever possible students can be challenged to try to think of at least two ways of sorting the same set of objects.

(cont'd next page ...)

The students themselves provide opportunities to find out things about themselves that are the same and different, without saying anything that might be hurtful. *Flexible thinking* and the idea that a problem can be solved in different ways are illustrated if a pre-selected group of students appears, one at a time, from the cloakroom and joins one of two pre-determined groups, i.e., same height and different heights. The remaining students try to guess how these students were sorted. Each time the students disappear and reappear again, different physical characteristics or characteristics of what they are wearing are used as sorting schemes that have to be identified. This is a suitable setting for students to get to know more about themselves.

The students are assigned to groups. The challenge to think of different ways of sorting the members of the group they belong to can be presented. The results are shared as riddles to be solved by the members of the class.

A group of six students goes into a cloak room and decides on a particular way of sorting the members of the group. The students come out and stand in the two groups. The class has to try and guess what it is. Points can be given if they either guess the actual sorting strategy or have one which can be applied. The points go to either the group or the class. Since everyone is part of both, everybody is rewarded.

After students have had experiences with sorting activities they are asked to try and think of when different people might use this strategy in their lives. A list that is prepared on their own or with a partner is shared with the other students. A riddle or pantomime setting may be suitable for a reporting task for some of the results.

Accommodating Responses

During sorting and classifying tasks it may be advantageous or necessary to challenge or remind students to try and think about characteristics other than colour, shape and size. As is the case with any group of students, there always will be unexpected responses from individuals. It was impossible to guess the strategy *'these are interesting and these are not'* used by one young student.

A letter from a former student who started his teaching career in a grade one classroom included the following anecdote,

> One student, when faced with several students at the front of the room and asked what was the same about them (initial classification experiences and all that) replied, 'They all don't have nothing (sic) the same about their hair colours.'

Out of the mouths of students! The conclusion included in the letter was, *Beautifully logical, you see, but now the challenge is to teach him something.* How true. It can be a challenge – an enjoyable challenge!

No doubt people who design tasks of the following type have one answer to the question in mind.

> *Which one does not belong?*
> [4]   [T]   [7]   [8]

It is easy to accommodate students who do not select the **T** as a response in a one on one setting. If necessary, redirection of questioning can be employed more than once by asking:

> *Which other one do you think could not belong? Why not?*

If this type of task is part of an activity sheet or part of an assessment presented to the whole group and is then marked, it can happen that choices other than the **T** are marked as incorrect.

Two possible scenarios exist. Incorrect choices can be based on logical thinking as the following responses to the question from students indicate:

> For selecting **8**: *'It is rounded.'* and, *'It has two insides.'*

Correct choices can be based on thinking other than what authors of items had in mind. One grade one student selected the **T** because,

> *'It is the only one with a roof.'*

Incorrect responses require follow-up questions and there may be times when the same is true for correct choices.

For some assessment tasks the instructions have to be very specific. For example, if it is the intent to find out whether or not students are able to distinguish between letters and names for numbers the instructions to students need to say so, i.e.,

> *Which of these is not a name for a number?*

Assessment Suggestions

The responses to open-ended questions and comments made during open-ended task discussions are a rich source of data related to:

- Indicators of *thinking flexibly*. Is a student able to consider characteristics other than colour, shape and size?

- Indicators of *risk taking*. How does a student respond to requests like, *Try to think of another difference for ...*; and, *Try to find another way of putting 'these' into groups that are in some way the same*.

- Indicators of *solving a problem* in more than one way: At least two ways of sorting the same objects or, volunteering two different responses to the sorting task request, *Which one do you think is different?*

- Indicators of *connecting*. Awareness of who uses sorting and classifying.

Sometimes the assessment reports about students who have been interviewed include the inappropriate descriptors 'seems' and 'very'. For example: *... seems confident* or, *... is very confident*.

Appropriate assessment statements should only make reference to what was observed during an interview or collected as part of an assessment strategy. Any comment that includes 'seems' is inappropriate. Nobody would accept a statement from a doctor that includes an observation like, *'You seem to be sick.'*

Any person who reads a statement that includes the descriptor 'very' has no idea about how many categories above or below this level an author of such a statement had in mind, nor about the criteria that were used to draw this conclusion.

One student shared a story about how his son who was in grade two kept choosing library books about science, his favourite subject, that were above his reading level. One evening he was struggling through a paragraph of a book mispronouncing many words. When he got to the end of the paragraph he concluded, *'Wow, this guy sure does not know how to write.'* Now that is an indicator of confidence!

Matching

The main goals of activities that involve matching include being able to find 'as many' or 'the same number of' objects without counting. Students need to learn that once objects in different sets are placed into one-to-one correspondence, the equivalence is maintained no matter what is done to the objects as long as objects are not added or taken away.

The language that is used when objects in sets do not match needs to be introduced. 'Big' or 'bigger' are not appropriate, even if objects in one set are bigger in some way. The appropriate terms are, 'more'; 'fewer'; 'not as many'; and 'not the same number.'

Students are asked to think of actions and incidents from their experience when and where matching is used. How do people talk about the results of this matching?

Types of Activities and Problems

- The classroom setting offers many opportunities to find 'as many' or to 'find the same number of' without counting. For example, students can help to solve problems like:
    - Find a pencil for each piece of paper.
    - Find a book for each desk.
    - Find a coat for each coat hanger.
    - Get a paint brush for each paint can.
    - Line up a boy for each girl that is lined up.
    - Select a block for each finger on one hand.
    - Assign a colour for each day of the week.

    The important follow-up question for each task is,
    > *How do you know there are as many of ... as there are ...?*
    The answers should make reference to ideas like,
    > *'There is one of ... for each one of ....'*, or, *'I matched them.'*

- Students need experiences with activities that involve gathering up the objects or members of one of the two sets whose objects have been matched. If students have concluded that there is the same number of pencils, or just as many pencils as there are pieces of paper, they could be faced with the scenario of looking at the pencils being picked up and held tightly in one bunch. If students agree that there are still as many pencils as there are pieces of paper they need to face the question,
    > *How do you know there are still as many pencils as there are pieces of paper?*

- The matching tasks assigned to students should include objects that differ in colour, shape and size. If, for example, twenty students stand around a table and each one gets one tiny block, it may not be easy for some of them to conclude that indeed, there are as many students as there are blocks. During the very early stages of learning some very young students might get distracted by the differences in size and be tempted to draw the conclusion about the students that,
    > *'This twenty is more.'*

Assessment Suggestions

The following questions can provide a focus for assessment data that can be collected about the understanding of one-to-one correspondence:

- Is a student able to find 'as many' without counting?

- Is a student able to explain why two sets whose objects have been matched have the same number?

- Does a student realize that rearranging the objects in one of two sets that have been matched, does not change the equivalence that had been established?

- Does a student use the correct comparison language for sets whose objects do not match?

- Is a student's understanding of matching not swayed by extreme differences in objects that are matched?

- Does a student recognize that number can be used to sort groups of objects?

- Is the student able to *connect* matching to actions outside the mathematics classroom?

What is Rational Counting?

The request, *Please count for me* is readily responded to by the majority of four and five year old children. They seem to enjoy the task. When they are asked to stop counting and then name the number that comes next, most of them are able to do so. However, when the question, *Why does that number come next?* is posed, some shrug their shoulders while other will respond with *'Because'*, often in a tone of voice that implies, *'do not ask me anymore questions about it.'* These children do not know the answer because they do not understand counting, but that does not stop some of them from providing a charming response. How is one to argue with rationales like, *'That is how it is.'* and, *'My mom told me.'*

The ability to count rationally or with understanding involves being able to look at ordered sequences of numbers; of numbers and numerals; or of numerals and:
- recognizing the growing patterns of the sequences,
- visualizing the numbers for numerals that are part of the sequences,
- visualizing the procedure of matching or one-to-one correspondence, i.e., match and add one more; or match and add two more; etc. to continue the sequences.

The ability to identify and describe hidden members of a growing pattern of numbers or numerals is also part of being able to count rationally.

Counting with understanding not only involves the *mental strategies* of *ordering* and *thinking about patterns*, but requires that students are able to *visualize* numbers.

Ordering

Ordering is a pre-requisite for learning to count rationally. This *mental strategy* is also part of other areas of *mathematical thinking* and learning. Students need to learn what is meant by ordering or putting things in order. The recognition of order will enable students to identify the characteristics used to establish the order and describe how adjacent members differ. Ordering tasks should require students to use a wide variety of characteristics as they extend ordered sequences, insert into them and construct their own.

The idea of *ordering* and the terminology can be introduced by randomly selecting several students and having them stand facing the remaining students. This selection should exclude the shortest and tallest students. The students are invited to guess what is being done as the students are arranged in order from tallest to shortest. Hints may be required and it may also be necessary to remind students to use descriptors other than 'big.' Students are told that this arranging is called, *ordering students according to height* or, *putting students in order from tallest to shortest*.

The ordering tasks that are part of the students' activities should make them aware of the fact that ordering can proceed in any direction.

As students participate in tasks that involve ordering, they should be challenged to try and think of incidents or events outside the mathematics classroom that involve ordering. Who is interested in putting things in order? Where? When? Why?

Types of Activities and Problems

- Three students are selected, one at a time, to extend the *ordered sequence* in both directions and to insert into the sequence. For each addition the students are asked to respond to,

  Where does this student belong?
  Why does this student belong there?
  How do two students who stand next to one another differ?

  The problem of inserting a student into the sequence who is as tall as a member of the sequence is presented,

  Where does this student belong?
  Why there?

  As part of a conclusion to this type of task, the students could be invited to respond to,

  You know how the students in the sequence are different from one another. Try to think of as many similarities as you can for all of these students. How are they all in some way the same? Make a list.

- Problems of constructing ordered sequences, extending and inserting into these can be created with a variety of materials. For example:
  - Glasses or bottles of the same size and shape filled with different amounts of water.
  - Pieces of paper with different shades of the same colour.
  - A set of bells of different sizes.
  - Dowels of the same length that differ in diameter.
  - Name tags with the same spacing between letters.
  - Sets of weights.
  - Pieces of paper that have the same shape, but differ in size.

- After students assist with ordering six name tags from shortest to longest, they are asked to suggest how six glasses that contain different amounts of water could be matched with the ordered sequence of name tags. It would be a very unusual experience, especially the first time this request is made, if students will not match the container with the least amount of water with the shortest name. The matching is continued until the glass with the most water is assigned to the longest name tag. Then the students are invited to react to the following requests:

  What if the glass with the most water is matched with the shortest name tag, which glass do you think should be matched to the next name tag, and the next one?
  What is the same about the two ordered sequences?
  What is different?

  There will be some young students who will suggest that, *'This can't be done.'* For these children establishing order and matching ordered sequences means that, *'Big things have to go with big things.'*

(cont'd next page ...)

■ Five sketches or cartoons depicting an action of some sort are presented. Examples of the action could be going and being on a fishing trip; taking a model airplane and launching it in a field; getting ready and going to a soccer game. The students are asked to identify the pictures they think should come first and last and to explain the reasons for their choices. After the remaining sketches are placed between those that were identified as first and last, the students are invited to tell a story for the ordered sequence of sketches.

The students are asked to consider whether another ordered sequence for the sketches might be possible. If there is a positive response, the suggestion is followed up with rearranging the pictures and listening to the matching story. If a suggestion is not forthcoming, the sketches are rearranged and the students are told that someone came up with this ordered sequence for the pictures. The students are challenged to try and come up with a meaningful story for the sequence of sketches. Rather than using sketches, five short sentences that describe actions could be presented to the students and the setting described for the sketches or cartoons could be replicated.

Questioning – Use of Imagination
The questions and requests that are part of initial ordering activities are straight forward or 'closed.' These types of questions can include:
  - *Put things in order from … to …*
  - *Where does this … belong in the sequence?*
    Why does it belong there?
  - *How are the things next to each other different?*

After the initial experiences with ordering tasks students could be presented with questions and requests that are open-ended. Attempts to satisfy such requests can stimulate students' imagination.

Pairs of students could be presented with five sticks or tongue depressors of different lengths or five pieces of cardboard that have the same shape but differ in size. After the request to put the objects in order has been met, several duplicates of the members of the ordered sequence and some that are longer and shorter, or bigger and smaller for the pieces of cardboard, are handed out. The students are requested to be ready to explain their thinking for any responses they generate for,
  What do you think could come next? And after that?

This request is open-ended and allows students who are flexible in their thinking to respond in different ways. For example, some may think of creating a repeating pattern while others may replicate the ordered sequence in a backwards fashion.

(cont'd next page …)

A direct invitation to students to *think flexibly* and to use their *imagination* could be extended by making the request,

> *Try to think of several different ways of answering the questions,*
> *What could come next? And after that?*

Fairness of questioning can be an issue for *ordering* tasks. The question, *What comes next?* for an ordered sequence on an activity sheet or as part of assessment can be answered in different ways by students who are able to think flexibly or who use their imagination.

Even if several choices for a possible answer are provided and students are requested to, *select the one that comes next*, different answers are possible. Assessment tasks require very specific questions or very specific instructions that remind students of what needs to be considered as they try to identify or describe the next member of an ordered sequence.

 # Assessment Suggestions

The following suggestions provide some guidance for collecting assessment data as observations are made while students are participating with tasks about ordering.

- Indicators of a student being able to describe what is meant by ordering or putting things in order.

- Examples of use of correct language as characteristics that are used to construct an ordered sequence are described.

- Examples of correctly making reference to the magnitude of a characteristic as objects are used to extend an ordered sequence and are inserted into an ordered sequence.

- Examples of being able to connect ordering to activities or incidents outside the mathematics classroom.

- Possible indicators of *confidence, risk taking or use of imagination* while:
  - reversing an ordered sequence.
  - dealing with an object that is the same as an object in the ordered sequence.
  - making suggestions for putting sketches or cartoons or sentences in different ordered sequences.
  - responding to open-ended requests.

Thinking about Patterns

Students need to learn to distinguish between designs, patterns, and designs with patterns. In non-mathematical settings, the term pattern is frequently used when design would be appropriate. When students are able to distinguish between design, pattern and design with a pattern, they could be challenged to draw three pictures of pieces of wallpaper or to create three flags to show how they think they would illustrate the difference for each one.

The students will learn to distinguish between repeating and growing patterns and will realize that one can easily be changed into the other and that there are many different ways to extend any given pattern.

Since growing patterns are a pre-requisite for counting rationally, they should be the focus of readiness activities. However, questions and problems about patterns can provide opportunities to foster *flexible thinking* in students and get them to use their *imagination*.

The initial activities should be designed to enable students to use their own words to define and describe what a pattern is. To reach this goal, a repeating pattern is displayed on the overhead or is drawn on the chalkboard.
For example:

paper clip   thumbtack   paper clip   thumbtack   paper clip   thumbtack

The students are invited to suggest what they think is special about this sequence of objects. After the students are asked to close their eyes, one of the objects is hidden behind a piece of paper. The students are requested to suggest what they think is hidden and to explain how they decided on the response.

The students are told that when it is possible to identify a hidden part or hidden parts of a sequence by looking at the other parts of that sequence, the sequence is called a *pattern*. If the same objects are used in the same way to extend the sequence, this is called a *repeating pattern*. The students are asked,
> *Who might be interested in repeating patterns?*
> *Has anyone seen repeating patterns? Where?*

Five circles in a row are shown on the chalkboard. Two dots are drawn, one above the other, into the first circle.

The students are shown that matching is used to enter two dots into the next circle and then two more are drawn above these dots. This method of, 'first matching and then two more' is demonstrated as dots are entered into the remaining circles. The students are told that if this procedure is continued the sequence is called a *growing pattern* because we can predict or tell what comes next.

The third plate in the sequence is covered and the students are requested to suggest at least one strategy for figuring out how many dots there are inside the circle that is hidden.

(cont'd next page ...)

After students have completed some activities with patterns, they could be asked to write their own definition for a pattern. How would they answer a younger student or a parent who asks them, *What is a pattern?*

The following comments come from a collection of definitions authored by students in grade two. A few of their definitions included phrases like: *match on and on*; and *keep on using the same shapes*.

Some students made sketches to show their definitions. When young students talk and write there will always be some responses that are worth a little chuckle:
> *A pattern is one shape after another.*
> *A pattern is different colours in a pattern.*
> *A pattern is very neat.*
> *A pattern is fun and good for you to practice.*

It is beneficial for students if the composing of their own definitions is revisited, even more than once. It takes time for mathematical language to develop and before students will include the key words and ideas that are part of definitions.

Purposes and Goals

Growing patterns are part of this chapter because these patterns are an integral part of *rational counting*. These patterns are also used by students when they develop *mental mathematics strategies* for some of the basic facts.

Some references for teachers include the suggestion that activities with patterns are part of pre-algebraic thinking. These references do not explain how this thinking is defined and how it is fostered by completing activities with patterns. It is rather doubtful that any of the activities with patterns that are included in the pupil texts for the primary grades transfer to any aspect of algebraic thinking. However, activities with patterns can be used to foster *flexible thinking* and *advance thinking* in young students. Appropriate tasks and questions can contribute to fostering *risk taking*, *confidence* and use of *imagination*.

Types of Activities and Problems

A partial and selective list of materials and settings illustrates that many possibilities exist for having students think about patterns:
- Sets of blocks of different shapes and the same colour for patterns with shapes.
- Sets of blocks of different shapes and colours for patterns with shapes; shapes and colours; and colours.
- Toy animals or cars.
- Pictures of animals or objects.
- The students in the classroom.
- Title pages of books.
- Body movements or shapes made with the body.
- Different instruments for rhythmic patterns with one instrument or a combination thereof.

(cont'd next page ...)

Accommodating Responses

Special care may be required when students are asked to repeat rhythmic or body movement patterns.

A correct repetition could be followed up with, *How did you know what to do?*

Different reasons for incorrect responses could exist.

A student may believe it was done correctly. In this case the response to, *How did you know what to do?* can provide possible insight into the information that was missed or is missing.

Since the repetition of rhythmic patterns requires auditory discrimination as well as auditory memory, the reason or reasons for not being able to repeat such a pattern may not become known. If a student realizes that a pattern was not copied correctly, asking *Why do you think that happened?* can yield information about the student's understanding.

The Power of Open-ended Tasks and Appropriate Requests

It was pointed out as part of the discussion about assessment questions for sorting and ordering activities that there exist types of questions that are not appropriate or are unfair for students, especially if responses are marked as being either right or wrong. The same dilemma exists for assessment questions about patterns.

On some activity sheets and tests students are shown a pattern and they are asked to respond to,
> *What comes next?*

This question is used for the following examples which are included on a page of one reference for students:

It is very unfair to students if the persons who use these items have one correct answer in mind. This is the case in some settings. Students who do not record the expected answers get their responses marked as being incorrect. These marks lead those who receive them to conclude that there exists a lack of knowledge about patterns. Nothing could be further from the truth.

One five-year old student who was asked, *What comes next?* as he was looking at the shapes: △ □ △ □

responded with, '*Circle.*' When asked to explain his thinking, he responded with, '*I want to see a circle.*' Fortunately the question, *What would you put down next?* was asked. The boy continued to build a repeating pattern that included the circle he wanted to see.

(cont'd next page ...)

The question, *What comes next?* is inappropriate if the intent is to have students extend a given patterns in a certain way. Since *repeating patterns* can be continued in many different ways and these patterns can easily be changed into *growing patterns*, or vice versa, many options exist for what can come next that satisfy the request of creating a pattern. Consider responses to the question for two of the patterns that were presented:

- students could think of a *growing pattern* and create the following:

Ann Bob Carin Ann Ann Bob Bob Carin Carin

- one more could added to each number to result in:

1   3   5   2   4   6   3   5   7   4   6   8

If the intent is to elicit one specific response, very detailed instructions are required. These instructions require careful attention and are not easy to design. Quite a bit of reading may be required by students. Some authors realize the complexity of the task of writing these instructions and they try to circumvent the problem by providing options and asking students to select from the choices. That does not solve the problem since there will be students who are so flexible in their thinking, they might be able to use any given option to create pattern.

As is the case for *ordering, flexible thinking, willingness to take risks* and *confidence* can be fostered in some students if the request, *What comes next?* is changed to,
> *What could come next?* or to,
> *What do you think could come next?*

The request, *Try to think of many different ways to create a pattern from a given pattern* gives students many options and provides an opportunity for use of *imagination*.

Changing a request from,
> *Try to find two different ways to extend the pattern*, to
> *Try to find at least two different ways to extend the pattern*,
will challenge some students to keep on trying and in this way individual differences are accommodated.

Documentation

A record of students' responses on activity sheets can be used to foster *flexible thinking* and to *advance thinking*. Activity sheets that show the different ways students were able to extend patterns are a valuable source since this documentation can be used to remind students of the ideas they had previously generated. The challenge of trying to go beyond what was done and to think of other ways can be presented.

From time to time, patterns can be revisited by presenting students with the challenge of trying to identify and describe members of repeating and growing patterns that are hidden.

Students are asked to explain the thinking they used to identify one or more missing parts of patterns and suggest how the patterns could be extended.

For example:

> *What do you think is missing? What do you think comes next?*
> *Explain your thinking.*

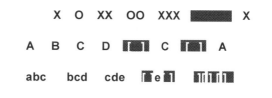

X   O   XX   OO   XXX   � ▮   X

A   B   C   D   ▮   C   ▮   A

abc   bcd   cde   ▮e▮   ▮▮▮▮

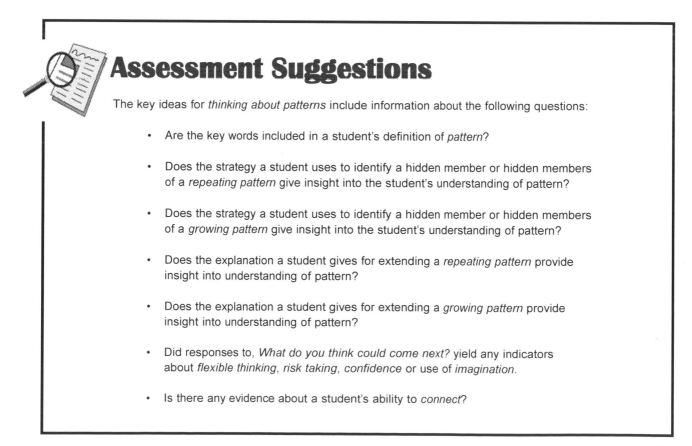

Assessment Suggestions

The key ideas for *thinking about patterns* include information about the following questions:

- Are the key words included in a student's definition of *pattern*?

- Does the strategy a student uses to identify a hidden member or hidden members of a *repeating pattern* give insight into the student's understanding of pattern?

- Does the strategy a student uses to identify a hidden member or hidden members of a *growing pattern* give insight into the student's understanding of pattern?

- Does the explanation a student gives for extending a *repeating pattern* provide insight into understanding of pattern?

- Does the explanation a student gives for extending a *growing pattern* provide insight into understanding of pattern?

- Did responses to, *What do you think could come next?* yield any indicators about *flexible thinking*, *risk taking*, *confidence* or use of *imagination*.

- Is there any evidence about a student's ability to *connect*?

Reporting

Sorting and classifying, matching, ordering, and thinking about patterns are essential for the understanding of number and counting. These strategies are an important part of all aspects of sense making in mathematics: *number sense, spatial sense, measurement sense, statistical sense* and *sense of relationships*. They are also part of other key areas of learning. Since that is the case, the important goals and learning outcomes should be shared with parents via a newsletter or during a meeting. This information will enable parents to make these strategies part of their conversations in and around the home

whenever the opportunities arise. These conversations can be a very valuable complement and supplement for what is learned in the mathematics classroom. Important ideas and correct usage of language can be reinforced.

The types of activities and problems that are part of teaching about the readiness strategies can yield valuable data about possible indicators of: *flexible thinking, confidence; willingness to take risks, use of imagination, use of appropriate language* and ability to *connect.*

Ideas for Parents

Sorting and Classifying
Conversations about similarities and differences should include as many different attributes as possible other than colour, shape and size.

It is easy to use parts of toys, objects from around the house or pictures to create settings that focus on the question, *Which one does not belong?* or, *Which one is different?* After the question is answered and a reason for giving the answer is explained, children should be invited to try and identify a second possible answer.

For a given group of objects like toy animals, toy cars, pictures of flowers, etc. children are invited to,
> Look at them and pick out those that are in some way the same.
> How are they the same?

After the objects have been sorted and they are returned to the original group, the child is challenged to,
> Try and think of another way of sorting these things.

(cont'd next page ...)

Matching

There exist opportunities in the home to get away from counting and create settings for matching tasks that involve the expressions 'as many' or 'the same number of' and then explaining the results.

As a table is set, the request is made to find, without counting, as many spoons as there are bowls or a cookie for each plate.

> *How do you know there are as many spoons as there are bowls*
> *or the same number of spoons and bowls?*
> *If the spoons are picked up and held in one hand, are there still*
> *as many spoons as there are bowls? Explain.*

Children need to learn to realize that once one-to-one correspondence has been established between the discrete members of sets, the number of the sets is the same no matter how extreme the difference in appearance and size and no matter how the members are arranged or rearranged.

When the discrete objects of sets can be put into one-to-one correspondence, we say there are, 'the same number of objects;' we do not say, 'they are the same size.' When the objects of two sets cannot be put into one-to-one correspondence, we say that, 'one set has more and the other has fewer objects'; we do not say, 'one set is bigger and the other has less objects' or 'one set is bigger and the other set is smaller.' The terms bigger, less and smaller are used for continuous quantity.

Matching can be used to identify or create sets that have the same number. If a child recognizes that two of five plates have the same number of cookies or blocks on them, number is used as a sorting and classification scheme.

Ordering

Most homes have toys of some sort that can be used for activities that involve putting things in order. These toys might include such things as cups, barrels, rings, Russian dolls or weights. If toys are not available, sticks of different lengths and pieces of cardboard that are the same shape, but differ in size can be used.

The major goal is to have children learn how to examine many different attributes and learn how to put things in order according to the magnitude of these attributes. As objects are put in order and as adjacent objects in an ordered sequence are compared and described, children learn to use the appropriate descriptors of the attributes rather than using the words '*small*' and '*big.*'

The first task requires the identification of the extremes:

> *Which cup can hold the least amount of water and which cup*
> *can hold the greatest amount of water?*
> *Which is the longest stick and which is the shortest stick?*

Then the remaining objects are ordered according to the attribute that is considered. After the objects are put in order from holding the least amount to holding the greatest amount, a few objects that were held back are introduced, one at a time. As children insert into and extend an ordered sequence, they are asked about how adjacent objects differ. Reminders may be required for some time to talk about something other than bigger or smaller.

After children have constructed an ordered sequence, i.e., the cups, they can be asked to do the same thing to another set of objects, i.e., a set of weights, and talk about how adjacent members in this ordered sequence differ.

Thinking about Patterns

The distinction needs to be made between designs, patterns and designs with patterns. By definition, pattern means that something is predictable. If a hidden member or members of a sequence can be identified by looking at the members that are not hidden, there exists a pattern. Any drawing or sketch without a pattern can be referred to as a design.

Whenever possible patterns should be identified and a conversation should take place that explains why a pattern is being looked at. Any objects from around the house can be used to create repeating or growing patterns.

Asking the question, *What comes next?* and having only one correct answer in mind is not appropriate for patterns since repeating patterns can be changed to growing patterns or vice versa. The only way to find out whether a child realizes that a specific pattern is being considered is to hide one or two members of the pattern and to find out whether or not the child can predict what the hidden objects look like by looking at the objects that are not hidden.

It is possible to find out something about a child's willingness to take risks by showing the child a pattern and making the open-ended requests,

> *Who do you think could come next?*
> > *Explain your thinking.*
> *Is there anything else that could come next?*
> > *Explain your thinking.*

For Reflection

 What would you say to parents who tell you that their young children know how to count because they sing counting songs, recite counting rhymes and count things whenever the opportunity arises?

 What comments would you make to a parent who makes the following statements about his son who just entered Kindergarten, *'He knows his numbers. He can count to thirty-five. Now he is learning his letters.'*

 What would you say to a parent who asks about the possible value of colouring books that have the young child count objects and then colour the objects and the numeral?

 How would you respond to a parent who asks, *'Why is it disadvantageous to tell my child what comes next in a pattern?'*

How would you respond to someone who states, *'I want my child to know the right way of doing things. I don't care if they can think of different ways to respond to questions.'*

Chapter 5 – Developing Number Sense

Number sense is the key foundation of numeracy.[1] Number sense is a pre-requisite for enabling students to develop:

- *mental mathematics strategies* for the *basic facts*,
- *personal strategies* for computational procedures,
- *mental mathematics strategies* for computational procedures, *and*
- *estimation* strategies.

Number sense is required for the development of *numerical power*.[2] The authors' list of characteristics of *numerical power* includes developing meanings of numbers and making sense of numerical and quantitative situations. The characteristics of classroom practices that promote *numerical power* identified by these authors reinforce the fact that the settings created by a teacher, the methods of orchestrating discussions and the accommodation of all types of students' responses are essential for the development of *number sense* and *numerical power*.

What is Number Sense?

The development of *number sense* requires special attention and time needs to be set aside for its development every year.[3] Whenever possible fostering this development needs to become part of ongoing mathematics teaching and learning as well.

The important aspects of *number sense* can be used to identify the major learning goals for students. These aspects include:

- ***Visualizing*** number: When students hear a number name or see a numeral, i.e., **five – 5**; **twelve – 12**; **twenty-four – 24**; etc., they will be able to 'see' the numbers for these names and numerals.

- ***Recognizing*** numbers without having to count, *subetizing*: A brief glance, too brief to use counting, will enable students to name the numbers that are shown.

- ***Flexible thinking*** about numbers: Students will show and name numbers in different ways. This ability is an essential pre-requisite for developing *mental mathematics* strategies and *personal strategies*.

- ***Estimating*** numbers: Students will use *referents* as part of their estimation strategies.

- ***Relating*** numbers: Students will use appropriate language as numbers and numerals are compared and ordered.

- ***Connecting***: Students will identify where numbers and numerals are used as part of everyday experiences.

The *mental strategies* of *sorting* and *matching* enable students to identify and select sets that contain the same number of objects. One group of discrete objects is used as a master set and after students have identified other sets that have as many discrete objects, the cardinal number is assigned to these sets and the symbol for the name is introduced.

Some student references use sets of dots of the same size as master sets, but it is easy to show that for the development of *number sense* the use of groups of fingers as master sets has many advantages.

The cardinal numbers are introduced to students in random order because later the *mental strategies* of *ordering* and *thinking about patterns* are applied to introduce counting with understanding or *counting rationally*.

Numbers and Numerals to Five

The fingers on one hand are well suited to introduce students to the first cardinal number. A request is made to get students to collect from containers in the room as many objects as there are fingers on one hand, to take these objects to their desks and place them onto a piece of paper. The students are to show, without counting, that there are as many objects as there are fingers on one hand.

As the students look at the groups of objects on other desks, they are asked to state anything they see that is different about these groups. Then the challenge to report all of the similarities they detect is presented. The observation that all sets have the same number of objects concludes with telling that whenever objects can be matched with the fingers on one hand, no matter what these objects look like, the same name is assigned to *that many* objects. The name **five** and the numeral **5** are recorded and introduced as names for this number or for 'this many' objects.

Types of Activities and Problems

■ Activities are designed to show students that five can look different as far as arrangements of objects and the size, shape and colour of these objects is concerned. The discussions following these types of activities focus on the question,

What is different about each way of showing the number five?

Tasks can include ideas like:
- The use of sketches to show different ways of decorating cookies with five raisins.
- Building different looking houses with five blocks of the same size and with blocks of different sizes.
- Designing different looking flower beds with five differently shaped pieces of coloured paper.
- Creating different looking designs with five toothpicks on different pieces of paper.

(cont'd next page ...)

The results of one of these tasks could be displayed on part of a bulletin board. The students can be challenged to assist with making up a title for the display, using five words about the number five.[4]

- Flexible thinking about number is developed when, with a partner, students are requested to come up with as many different arrangements as possible to show five with fingers on both hands. For each arrangement they are to show, without counting, that there are five fingers. This is done by matching the fingers for each arrangement with the fingers on one hand.

- Arrangements of five objects or things are prepared while the students were not in the room, i.e., designs with thumbtacks; groups of words and letters; paint cans; pictures; etc. As part of a *Hunt for Five*, students are asked to look around the room for fives and to be ready to show without counting that what they see shows five. Then the students are asked to try and find groups that do not have five objects, to show without counting that there are not five, and to verbalize a comparison of the master set and the objects in the other sets using the words *more* or *fewer*.

- Objects are placed on an overhead projector or dots are drawn on one-half of a piece of cardboard. The students are asked to look at the fingers on one hand as they watch the screen or the folded piece of cardboard. The request is made to use a thumb-up signal if they think they see more than five objects on the screen or the piece of cardboard, and a thumb-down signal if they think they see fewer than five objects. The overhead is briefly turned on or the cardboard briefly unfolded. This task gives students the idea of using a *referent* as a prediction or an *estimate* is made about number.

- A name tag could briefly be shown to students and they could be asked whether they thought there were more or fewer than five letters in the name. The students should be ready to explain how they arrived at the answer or estimate.

Similar types of activities are used for introducing the numbers and numerals for four and three.

One very observant boy in grade one who had been part of these types of tasks declared, '*You can't do that to two.*' This conclusion made reference to the request of showing different arrangements for the same number.

Student are told that *zero* is the name assigned to the number of fingers that are shown when no fingers are held up.

For one task that fosters *visualization* fewer than five fingers are briefly flashed. The students are asked to provide answers for,
> *How many fingers did you see?*
> *How many do you think you did not see?*

As the numbers and numerals are presented in random order, students are faced with activities that foster the development of the important aspects of *number sense*. Since the types of activities would be similar for each number, only selected activities for the number six are described.

Types of Activities and Problems

- The master set of six fingers, five on one hand and one on the other, is assigned the number name **six** and the symbol **6** is introduced. The students are told that these number names are assigned to any set of discrete objects that can be matched or put into one-to-one correspondence with that many fingers. Since differences in appearances for the objects in these sets can be extreme as far as size, shape colour and the arrangements of the objects is concerned, students could be invited to think of and describe sets of six objects that can differ greatly in appearance.

- Students are asked to select six blocks and place these onto their desks. Several students are asked to show without counting and to explain the procedure they use to show that there are six blocks in the collection of blocks. The request is made to use the blocks and build a house that looks a little different from the houses on adjacent desks.

  The follow-up questions are,
  > How are the houses on the desks different?
  > What is the same about all of the houses?

  The students are requested to hold up fingers to show the answers to the following questions and to explain their thinking:

  - *If each block or room had one window, how many windows would there be in each house?*

  - *If one person lives in each block or room of the house, how many persons would there be in each house?*

  - *If pets were not allowed in any of the houses, how many pets would there be?*

  - *If each room had one chair in it, how many chairs would there be in each house?*

  - *Explain your thinking for: Would there be more chairs or more windows, or would the number of chairs be the same as the number of windows for each house?*

- Two important aspects of *number sense* are **visualization** of number and **flexible thinking** about numbers. Types of activities that can foster the development of these aspects include:

- Six chocolate candies (dots) are to be used to decorate two cookies (ovals). How could the candies be placed onto the cookies? What are the possible ways?

- Think of as many different ways as you can to show six with fingers. Keep a record of the different arrangements with the word 'and', i.e., five and one. Do you think 'five and one' is different from 'one and five'? Why or why not?

- Six stars are shown on an overhead or on the chalkboard. A cloud or piece of cardboard rushes by and covers part of the stars. For each covering, show with your fingers on one hand how many you think are hidden and with fingers on the other hand how many are not hidden. How do you know your answer is correct?

- Different looking drawings of six sided figures are displayed. How are the figures different? How are the figures the same?

- Pairs of students are invited to prepare a list of all of the places, settings and actions where six is used or appears. The lists are shared and compared.

- The request is made to collect a list of names for the number six from different languages. How are the names different? Is there anything the same about the names?

■ After the numbers and numerals up to and including nine are introduced, activities that further the development of: *recognition of number; relating* numbers and numerals; and *flexible thinking* about numbers are presented.

As two students face each other, one briefly flashes a number of fingers, long enough to be seen, but too brief to be counted by the other student whose task it is to name the number that was shown and then to show it in a different way. This *recognition* and *flexible thinking* is an important pre-requisite for developing *mental mathematics strategies* for the basic facts.

As part of flashing numbers to each other, one girl in grade one showed ten fingers to her partner. The partner looked at her in a very puzzled way and as she held up ten fingers shrugged her shoulders to indicate that she was puzzled about showing this ten in a different way. This bewilderment was resolved when her wrists were taken hold of and the forearms were crossed for her. The problem was solved!

■ After a number of fingers are briefly flashed, a partner names the number, tells how many more fingers it would take to show ten fingers, shows the number in a different way and names the new number using the word '*and*.' For example, for six fingers flashed with the five fingers on one hand and one more, the responses would be:

- for number *recognition* or *subetizing*: '*six.*'
- for *visualization* of going to ten: '*four.*'
- for *flexible thinking* about number: showing two threes and stating, '*three and three.*'

- **Relating**
  A student is asked to stand in front of the class and use fingers to show the number seven. As the following requests are made, the students are asked to respond by showing fingers:
  - Show a number very *close* to seven.
  - Show a different number that is just as *close*.
  - Show a number *less than* seven.
  - Show a number *greater than* seven.
  - Show a number *between* seven and four.
  - Show a different number *between* seven and four.
  - Show a number that is *not between* seven and nine.
  - Show a number *less than* seven that is *far away* from seven.

- The different words used to make comparisons are recorded on a chart. Students are assigned a partner. Turns are taken to hold up a number of fingers and inviting the partner to respond to requests that use words from the chart.

- **Connecting**
  Throughout a school day, many questions are asked that require a number or number name as an answer. Whenever that is the case, students can be asked to show their answer by holding up the appropriate number of fingers and then showing that number in at least one other way.

Printing Numerals and Number Names – Reversals
Since the development of *number sense* and the ability to count rationally does not depend on the ability to print and write numerals, learning how to write the names and symbols should be part of printing lessons.

Quite a few young students reverse numerals or lack the ability to print or write numerals. This is not an indicator of lack of *number sense*. For the majority of students reversal difficulties will disappear over time. Some students can be helped with activities such as reversing on purpose.

Rational Counting

The major goal for counting with understanding is to enable students to answer the questions,
> *What comes next? Why?* or, *How do you know?*
> *What is hidden? How do you know?*

for any counting or skip counting sequence. This ability requires that students need to be able to *visualize* the numbers for counting sequences with numerals. To reach this goal, counting tasks have to begin with numbers before moving on to sequences with numbers and numerals, and finally to sequences with numerals.

Ordering Numbers
The mental strategies of *ordering*, *matching* and *thinking about patterns* are part of learning how to count with understanding.

As part of an introduction, students could be asked to order a set of numbers from least to greatest. After the task is completed an open-ended request is presented.

(cont'd next page ...)

For example: ■■ ■■■ ■■■■ ■■■■■

What number do you think could come next?

After different responses are discussed and compared, the students are told that if the next number is obtained by matching the last number of the sequence and then adding one more, the *growing pattern* is a counting sequence.
If this procedure is repeated, what would the next number be?

The students could be invited to respond to,
Do you think it would be possible to get to the greatest number?
Why or why not?

Ordering Numbers and Numerals
The next step in the teaching sequence involves the extension of counting sequences and labeling each number of the sequence with number names.

For example,
Think of a growing pattern and a counting sequence, what comes next?
Show the number and print the number name.

   1   3   5

Ordering Numerals
The last step includes extending numeral counting sequences in either direction, identifying hidden members and justifying the responses.

For example,

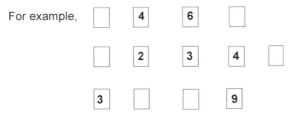

Visualization of numbers will be fostered and rote recital will be avoided if the same steps of presentation are adopted for any skip counting sequence.

Responses collected during diagnostic interviews include comments from students in grade one and two indicating that a growing pattern like: 2 4 6 ▢ is not possible because, *'there have to be numbers between the numbers.'* It is very likely that these students missed out on building patterns with numbers, then with numbers and numerals and finally with numerals. Moving too quickly to the symbolic level can result in lack of ability to *visualize* and being unable to 'see' the numbers connected to the number names.

(cont'd next page ...)

As part of a demonstration, objects on an overhead projector or on the chalkboard are counted incorrectly. The students are invited to watch and to raise a hand every time they think a mistake has been made. Errors in counting can include counting spaces between objects, counting an object twice, skipping an object, skipping a number name, continuing counting beyond the last object, or repeating a number name. After the demonstration students are invited to help with making up a rule for counting correctly.

Even and Odd Numbers

Fifteen counters of the same size and shape and three pieces of paper are provided for each student or for a pair of students. The request is made to place four, five and six counters onto the three pieces of paper, respectively. The students are invited to try and make different designs and try different arrangements with each group of counters. An open-ended question is posed,

> *Did you notice anything about these numbers?*

If the question is too open-ended, the students are asked to try and compare the numbers four and six to the number five.

> *Is there something that can be done with four and six counters that cannot be done with five counters?*

More guidance may be required for some students to lead them to conclude that some numbers can be arranged in two rows with the same number of counters in each. We call these *even* numbers.

> *Look at your fingers. How can you show that ten is an even number?*
> *Show with your fingers that nine* (seven; five; three) *are not even numbers.*
> *Use counters or draw dots to show eight different even numbers. Record the names for the even numbers. How is it possible to recognize a name for an even number?*

Numbers that are not *even* are called *odd* numbers.

Ordinals

When young students count a group of objects by rote, they are assigning ordinals to these objects. For example, they say '*five*' as they point to the fifth object. They are not visualizing all of the objects that have been pointed to previously. For these students it is accidental that if, by chance, number names are recited in the correct order as one object is pointed to, the last uttered name tells them the answer to, How many? *Rational counting* implies *visualizing* the cardinal number of the set that is counted and knowing how it differs from the previous number as well as the next number.

For an introduction to *ordinals* several students are invited to come to the front of the room, stand in a line and face in the same direction. The following types of questions are used to introduce ordinals,

> *Who is number one or who is first in the line?*
> *Who is number two or who is second in the line?*

Types of Activities and Problems

Tasks with ordinals can include:

- Ordering within a line-up or sequence. Who is the second boy or second girl? Who is the second boy with blond hair?

- Discovering that in certain line-ups of people or objects, one person or object keeps the same position when they face in the opposite direction or when the order is reversed. Students are invited to test the discovery for line-ups with different numbers of students.

- Recognizing that ordering is independent of direction. Ordering is very different from reading and writing; it can begin anywhere and can go in any direction.

- Coordinates: Labelling rows or tables as first, second, third, etc. and asking for responses to requests like:

  Would the second boy in the third row please raise his hand.

  Would the third girl in the second row please stand and take a bow.

  Would the second girl in the first row please wave at everyone.

- Notation: Tell students that for (**1,2**), the first numeral identifies the row and the second the seat in the row. While coordinates are recorded on the chalkboard, requests like the following are made,

  *Would (**1,4**) please print his name on the chalkboard.*

  *Would (**4,1**) please open the window.*

- Connecting: Identifying ordinals in settings or events outside the classroom.
  Who uses number names like these? Where? When?

Assessment Suggestions

As part of assessing the ability to *count rationally*, indicators are collected about being able to explain what comes next and why:

- Extending *growing patterns* with numbers. Specific instructions are required for this task. As is the case for *ordering* and *thinking about patterns*, the question *What comes next?* is not sufficient.

- Naming a hidden member of a *growing pattern* with numbers and explaining the thinking.

- Extending *growing patterns* with numerals. Specific instructions are required for this task. The question, *What comes next?* is not sufficient.

- Naming a hidden member of a *growing pattern* with numerals and explaining the thinking.

- Identifying numbers as *even* and *odd*.

- Identifying numerals as names for *even* numbers and explaining why that is the case.

- *Connecting*: Examples of statements that indicate when and where numbers and numerals are used.

As data about counting with understanding are collected, any indicators of *connecting*, *confidence* and *risk taking* are noted.

Reporting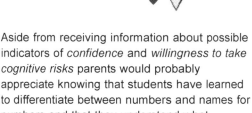

Aside from receiving information about possible indicators of *confidence* and *willingness to take cognitive risks* parents would probably appreciate knowing that students have learned to differentiate between numbers and names for numbers and that they understand what counting is.

Simple examples can be used to illustrate this new knowledge to parents. When students answer the questions *How many?* or, *How old are you?* with the help of their fingers, the students show numbers. The letters and the symbols they print on pieces of paper are names for these numbers.

When students are shown a partial counting sequence with a missing number name, or with missing number names,

they can explain why the answers they give for the following questions are the correct answers by making reference to the pattern of matching and one more or one less,

What number comes before five?

What number comes after eight?

What number is between six and eight?

The focus of the suggestions for strategies, questioning and activities is on fostering the development of *number sense* and accommodating its components: *visualizing, recognizing, thinking flexibly, estimating, relating* and *connecting*.

Types of Activities and Problems

Visualizing

Students need to be shown that **10** is a special numeral. When students hear the word '*ten*', it is very likely that they *visualize* ten discrete objects. A request to try and show and record all of the different ways ten name tags can be placed onto two covers of books will show that **ten** can have different names:

9 and **1**;   **8** and **2**;   **7** and **3**;   **6** and **4**;   **5** and **5**;   **10** and **0**.

As one child displays ten fingers, **10** is recorded on the chalkboard. The students are told that there is a special way to think about this number name. People say 'ten', but it would be more appropriate if they would say, '*one ten zero*' and think of it as, '*one person holding up all the fingers*' or as, '*one group of ten.*'

Rote learning about number names can lead to confusion, especially for some of the 'teen' numerals. Reversal problems can occur for some students. They will record 31 when 13 is called for or will read 13 as '*thirty one.*' These types of reversals are indicators of lack of *visualization*. The reading of two-digit numerals in two ways can foster the ability to *visualize*, can help to prevent these types of reversal problems and can contribute to the development of *mental mathematic strategies* for the basic facts. The request is made to read **13** in two different ways:

> '*thirteen*' and '*one ten and three.*'

Similar requests are made for other 'teen' numbers and number names.

Students are asked to look at the **15** in the sequence below and respond to,

> *Why is 'one ten five' a better name than 'one five' for this number name?*

The students are told they are looking at a counting sequence and they are asked to try and identify the number names that they think are on the backside of the pieces of cardboard:

Explain your thinking for the answers.

After the missing number names are identified and the sequence is extended to **11** and **19**, students are asked to compare this sequence to the counting sequence from **1** to **9**.

> *What is the same and what is different about the two sequences?*

(cont'd next page ...)

As students look at **20**, they are invited to respond to,
> *Name the fewest number of children it would take to show this number with fingers.*
> *What would be a good way to read this number name that would tell young children what it means? Why is that the case?*

Flexible Thinking

As two students face the group, one is asked to hold up ten fingers and the other three. The open-ended request is made,
> *Tell everything you can about the number you see.*

It may be necessary to provide hints about looking at the fingers on both hands and at the fingers and thumbs before students may recognize and report:
- two hands and three or, **5** and **5** and **3**;
- two thumbs and eleven fingers or, **11** and **2**;
- one hand and eight fingers or, **5** and **8**;
- one student or one group of ten and three more or, **10** and **3**.

All of these are different names for **13.**

Looking at the arrangements of fingers can foster *visualization*. This *visualization* and *flexible thinking* about the 'teen' numbers and numerals transfers to and facilitates learning about the *basic facts*.

Relating

When the discrete objects of different sets are compared, the terms *fewer*, *more* and *as many* or *the same number of __* are used. When numbers are compared, the terms *greater than* and *less than* are used.

Solving and Writing Mysteries

To find a hidden penny, cardboard cards or envelopes labelled **1** to **20** are displayed in order on the ledge of the chalkboard. A penny is taped to the back of the piece of cardboard labelled 11 or placed inside the envelope showing the number name **11**. Hints written on strips of paper or cardboard are shown to the students, one at a time, and placed into a display chart. After each hint, the students are asked to respond to,
> *Which number name or names might hide the penny?*
> *Which number name (or names) can be turned over because it is known that it does not (they do not) hide the penny?*

The number name that hides the penny:
- **Does not have a five in it.**
- **Is between two and eighteen. What is known about 2 and 18?**
- **Is not between six and ten. What is known about 6 and 10? Why?**
- **Is not between twelve and fifteen. What is known about 12 and 15?**
- **Is a name for a number less than seventeen.**
- **Is a name for a number greater than four.**
- **Is very close to ten.**

(cont'd next page ...)

Opportunities for reflection that involve *relating* are presented when students are asked to write or make up hints about a mystery number or number name. This should be done after they have had several opportunities to participate in tasks like the one about a mystery penny. Initial attempts should be for numerals **1** to **10**. Students are requested to try and use the words from a chart that lists terms that have been used as part of relating numbers and number names.

The following are first attempts from three students half way through grade one:

Student one: *it is not **6***
               *it is higher then **4***
               it is not one

Student two: *it is clos* (sic) *to **5***
               it is clos (sic) *to **6***
               it is clos (sic) *to **4** what is it*

Student Three: *it is higher then* (sic) ***4***
                it is lower then (sic) ***9***
                *it is not **6***

Students appear proud when they share their hints with their classmates. When invited to write another hint, most students will do so, but a dilemma may exist. They know the number name they have written about, and some find it difficult to imagine why others, especially the adults in the room, would not know what that number name is.

If students are to learn how to *communicate mathematically*, they must be given opportunities to do so. It will take a while before students use more words and before they make their hints more specific. Observations over a five week period, at different grade levels, have shown that to be the case.

Parents who find themselves in a classroom while students make their first attempts to write about their mystery number need to be educated not to finish sentences for students, as some will be and are tempted to do. The time it takes to learn how to write about numbers is not shortened by having someone do the writing and thinking for students.

It is always enjoyable and refreshing to encounter young students who are full of confidence, are flexible thinkers and solve problem in unique ways.

In one grade one classroom, the students were given the task of printing several number names and then printing a few hints about one of these. One young boy, who did not know how to print, was assigned to a parent helper. On his sheet appeared the numerals **1**, **2**, **3**, 4, 5, **6**, and **10**. The underlined numerals were reversed. The first three hints that were dictated to and recorded by the parent were:
       It is close to 4.   **It** (sic) **somewhere below 10.**   **It is not 2.**

It should be kept in mind that the students who record their own hints, can look back at these and refresh their memories about what they have stated. This boy could not read the parent's handwriting. The inability to retrace his thinking or reread his hints did not stop him from providing the next hint. After selecting a purple crayon and colouring beside the **5**, he dictated,
            It has a coloured line beside it.
An excellent example of a student solving a problem and knowing how to get unstuck!

(cont'd next page ...)

As the grade one teacher walked into the staffroom, she smiled. Her students had been given the task of printing true statements by inserting *is greater than* or *is less than* between two names for numbers. One boy was absent when *is less than* was introduced. This absence did not stop him from printing a true statement.

For the example, **2** _____ **4,** he had flipped the page, printed *is greater than* between the backwards **4** and **2**, flipped the page back, and traced the backwards printing of *is greater than* between the **4** and the **2**. This is a powerful example of confidence and problem solving.

Estimating and Recognizing

Students are asked to look at the fingers on both hands as objects on an overhead projector or dots drawn on one-half of a folded sheet of paper are briefly shown. Ten becomes a new *referent* for making decisions.

There exist many opportunities to use ten as a referent. As pictures of objects or animals are examined or as books on a shelf are pointed out, the request can be made,
> *Thumbs up if you think there are more than ten or thumbs down if you think there are fewer than ten. Explain your thinking.*

Experiences of this type will lead to the ability to have students *recognize* in many instances whether there are *fewer than ten*, *about ten* or *more than ten* members in a set.

Numbers and Numerals to Ninety-Nine

Students are told that to find the answer to the question, *How many?* the objects are grouped by tens as often as possible. A record is kept of the groups of ten or *tens*, and how many objects are left over or, the left over *ones*.

Types of Activities and Problems

■ *Visualizing*

One major goal is for students to *visualize* numbers when they see a two-digit numeral or hear a name for a two-digit numeral. The counting and grouping by tens is illustrated with forty-two toothpicks and elastic bands. Reading **42** as *'four tens and two ones'* and as *'forty-two'* contributes to the ability to *visualize* the meaning of the digits in the numeral.

Visualization is enhanced when students are asked to think of the fewest number of students it would take to show forty-two with fingers. The answer *'five'* is illustrated by having four students show the groups of ten fingers and one student two fingers.

The students are invited to respond to,
> *What is the greatest number of students needed to show forty-two?*
> *What are some other possible ways of showing forty-two with fingers?*

(cont'd next page ...)

- **35** is printed on the chalkboard. Students working with a partner are requested to record answers for:
    The least number of students it takes to show this number with fingers.
    If one more student joined the group of students showing thirty-five with fingers:
    - *What is the greatest number the students could show now?*
    - *What is the least number the students could show now?*
    Did any students think of zero as a number?
    - *What other numbers could these students show?*
    Record all the number names you can think of and order them from least to greatest.

- Three students are asked to show the **24** recorded on the chalkboard with fingers. The responses to questions of the following type are illustrated with the appropriate number of fingers and students.
    - *What number is ten greater than twenty-four?*
    - *What number is ten less than twenty-four?*
    - *What number is one more than twenty-four?*
    - *What number is one less than twenty-four?*
    - *What number is double or twice twenty-four? Explain your thinking.*
    - *What number is one-half of twenty-four? Explain your thinking.*

    These types of tasks are repeated with other numerals, some with **0** and **9** in the ones place.

- **Coins - Dimes and Pennies**: The students are to think of using only dimes and pennies for a given amount of money, i.e., **43 cents**, and supply the answers to,
    - *What is the least number of coins required to show the amount? Name the coins.*
    - *What is the greatest number of coins that can be used to show that amount? Name the coins.*

Flexible Thinking

Flexible thinking about numbers is a pre-requisite for developing *personal strategies* for computational procedures. Students need to realize that a number can be represented in different ways and can have different names.

As three students face their classmates, they are requested to show the number twenty-six with their fingers. The answer to *How many*? is elicited and recorded on the chalkboard, **26**. The numeral is read in two ways, as *two tens six* and *twenty-six*.

The request is made to examine the arrangements of fingers, thumbs and hands.
What other possible names for twenty-six do the arrangements of fingers show?

For example,
- five hands and one finger - **5** and **5** and **5** and **5** and **5** and **1**
- five thumbs and twenty-one fingers – **5** and **21**

If one student holds up ten fingers, how many students each holding up one finger would it take to show twenty-six? **10** and **sixteen ones** or students.

All of these show different names for twenty-six.

Similar tasks are presented for other numerals and numbers.

(cont'd next page ...)

In some references it is incorrectly suggested that **2** tens **6** ones or **20 + 6** are different names for **26**.

Tasks with Coins

- **Dimes and Pennies**: The students are to think of using only dimes and pennies for a given amount of money, i.e., **32 cents**, and use sketches to show the amount in at least three different ways.

- **Dimes, Nickels and Pennies**: Students are requested to select at least one from each of the coins and use sketches to show a given amount of money, i.e. **25 cents,** with the greatest number of coins, the least number of coins, and in two other possible ways.

- **Quarters, Dimes, Nickels and Pennies**: The request is made to select at least one of each of the coins and to show a given amount of money, i.e., **52 cents**, in at least two different ways.

- **Receiving Change**: Sketch at least three different ways of receiving change for a given amount of money, i.e., **37 cents**. Why are shop keepers interested in giving change with the fewest number of coins?

- **Riddles**: The students are told how many coins are used to represent a given amount of money. What are the coins?

  20 cents - 2 coins   **20 cents** – 4 coins
  20 cents – 8 coins   **20 cents** – 11 coins

Estimating

When students are introduced to estimation, they use the *referents* five and ten as they look at sets of discrete objects and report whether they think there are more or fewer than five or ten. The goal now is to have students look at a set of discrete objects of the same size and shape, use ten fingers as a *referent*, pretend to group the objects into tens, and report an estimate in the form of *'about _ tens.'*

- Early estimation experiences ask students to use ten fingers as a *referent*, pretend to group a set of discrete objects shown on an overhead projector or dots drawn on a piece of paper, and to choose an estimate from one of three or from one of two options.

  For example,
     Thirty-two chips, pennies or dots are displayed.

  Three possible choices for estimates could be:
     About ten or about 10
     About three tens or about 30
     About six tens or about 60

  Two possible choices for estimates could be:
     About three tens or about 30
     About six tens or about 60

  As students gain more experience, the range of the choices is narrowed.

(cont'd next page ...)

■ A different type of estimation tasks involves showing students two pictures of groups of people or two paper plates with different numbers of counters, i.e., **38** and **67**, respectively.

The students are asked to pretend to be grouping these people or counters into groups of tens.
  Which picture or plate do you think shows about four tens or about forty?
After one picture or plate has been chosen,
  If that picture or plate shows about four tens or about forty, about how many tens do you think there are in the other picture or on the other plate? Explain your thinking.

■ The word '*about*' is frequently used during everyday conversations that make reference to numbers. For example,
  We have about twenty kilometres left to drive before we get to the next town.
  The salmon I caught weighed about thirty kilograms.
  There were about one hundred people at the soccer game.
Students need to learn how to *visualize* the range of possible numbers that go with each statement that uses the term *about*.

Forty-two counters are placed into an envelope. The students are told that,
  There are about forty counters in the envelope.
The students are invited to respond to requests of the following type:
  • *Record two number names that you know do not tell how many counters there are in the envelope. Explain your thinking.*
  • *How do you know that there are not forty counters in the envelope?*
  • *Record two different guesses for the number of counters that you think could be in the envelope. How did you decide on these guesses?*

The different guesses the students report are recorded in order from least to greatest on the chalkboard. For any missing number names between **35** and **45**, the students are asked, *could there be 'this' many counters in the envelope? Why or why not?* The students are asked,
  Why is thirty-two an incorrect guess?
  Why is forty-eight an incorrect guess?
  What is special about thirty-five?

The students are told that a number in the middle is usually taken to the next highest ten. Some people may agree with this rule, but some may not like it.
  How do you think the following people would feel about this procedure? Explain your thinking.
  • Someone caught a fish weighing **35** kilograms.
  • A mother whose age is **35**.
  • A family that has to travel another **35** kilometres to get to the lake and everyone is anxious to get here.
  • You have **35** cents in your piggy bank.

Relating Numbers and Numerals

The abilities to *compare* and to *order* numbers and numerals are an important aspect of *number sense*. The vocabulary list that was used for comparisons of numbers and numerals to twenty can be expanded to include odd, even, ones place, tens place, nearest ten and digit.

(cont'd next page ...)

Solving and Writing Mysteries

- For an example of a *Mystery Number Name* setting, the numerals from **70** to **99** are printed on the chalkboard or they are highlighted on a **99**-Chart. After each hint, students are requested to:
  - *Name several examples that could be the Mystery Number Name.*
  - *Name at least one that could not be the Mystery Number Name. Explain your thinking.* Responses are crossed out.
    > **The numeral does not have a 9 in the ones place.**
    > **It is a name for a number greater than 72.**
    > **It is not between 72 and 77.**
    > **It is not between 91 and 96.**
    > **It does not have a 1 or 2 in the ones place.**
    > **It is a name for an odd number.**
    > **The digits in the ones place and the tens place are
    > not the same.**
    > **It is close to 80.**

- It is a valuable experience for students to examine statements that are not true. For the same range of numerals from the **99**-Chart as the last example, the students are told that this time each statement is not true. After each hint, the students are requested to explain their thinking for the answers to the following questions,
  - *If the hint would be true, what answer or answers are possible and what answers are not possible?*
  - *Since the hint is not true, what answers are not possible and what answer or answers are possible?*
    > **It is a name for a number less than 75.**
    > **It is a name for a number greater than 95.**
    > **It is between 89 and 95.**
    > **It is between 75 and 80.**
    > **It is a name for an odd number.**
    > **It is close to 80.**
    > **It is not close to 90.**

- After several examples of identifying *Mystery Number Names*, students are given the opportunity to author their own hints about an entry from the **99**-Chart. Some students might enjoy the challenge of trying to write hints about a *Mystery Number Name* that are not true.

- As students look at the **99**-Chart, an open-ended request is made,
  > *What are some of the things you notice about the number
  > names on the chart?*
  The request is made more specific by asking,
  > *If you see any patterns, what are they?*
  More guidance is provided when students are asked to compare the entries in different rows and different columns.

- Different number names and different sequences of number names on the **99**-Chart are hidden. The students are requested to try and think of at least two ways of determining what they think the missing number names are.

(cont'd next page ...)

■ Pieces from the **99-Chart** with missing number names are presented. The students are requested to suggest two ways for determining the missing number names.

Try to find two different ways of explaining how to find the missing number names.

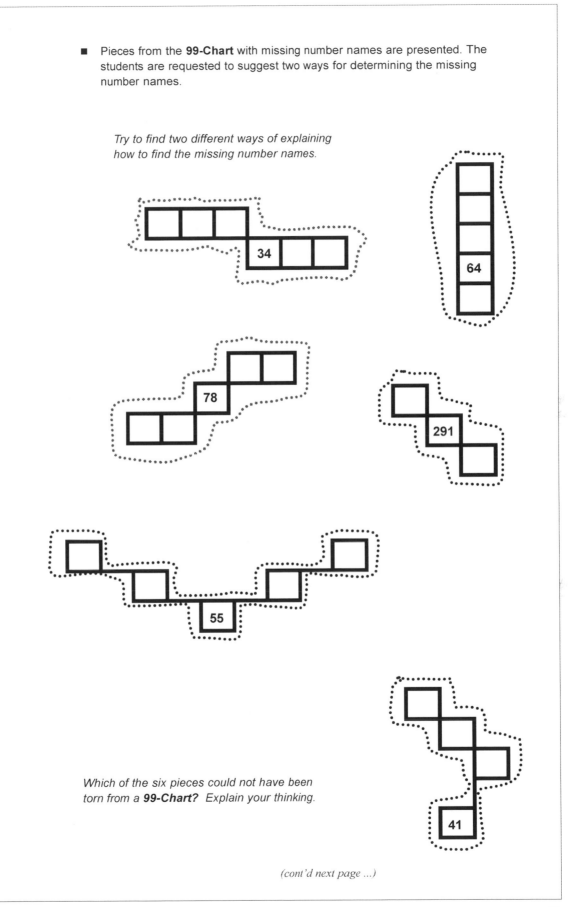

*Which of the six pieces could not have been torn from a **99-Chart**? Explain your thinking.*

(cont'd next page ...)

100 – a Special Numeral: Students are called to the front of the room and they are asked to hold up ten fingers. As groups of ten are displayed, students skip count by tens to announce the total number of fingers that are displayed. After the tenth student has come to the front and 'one hundred' is announced, the ten students are requested to stand shoulder to shoulder in a tight circle and to raise their hands above their heads.

As **100** is recorded, the students are told that there is a special way to think about these ten groups of tens. People think of the number that is shown as **one group of one hundred** and zero groups of ten and no ones.

Assessment Suggestions

Since the development of *number sense* is a key aspect of mathematics learning, assessment data about indicators of this development need to be collected and shared with parents.

It must be kept in mind that the development of *number sense* is a delicate cognitive process that takes place over a period of time. Any statement about students having *'good number sense'* is inappropriate because it is subjective and therefore meaningless to a reader. The best that can be reported at any one time is something about the indicators of developing *number sense* that have been observed and noted.

As students talk, write and report about numbers and number names, it is possible to note indicators of:

- **Visualizing**: Examples of thinking about the numbers as number names are discussed or written about.

- **Flexible Thinking**: Examples of realizing that numbers can be represented and named in different ways.

- **Estimating**: Examples of explaining how five or ten is used as a *referent* to arrive at an estimate about numbers or pretending to arrange objects into groups of tens to arrive at answers in the form of: *About _ tens*.

  There exist references about assessment rubrics that suggest that students' estimates can be classified into such categories as 'reasonable'; 'unreasonable'; or even as 'logical' or 'illogical'. These descriptors do not make sense.

  The only way to probe the ability to estimate is to have students explain orally or in writing the strategy that was employed to arrive at an estimate. If students use an appropriate estimation strategy, no matter what the results, they have to be accepted because it is 'logical' according to the students and their *sense of number* at that moment in time.

(cont'd next page ...)

A response may reveal something about development of *number sense* with respect to ability or inability to *visualize* numbers. Since the ability to *visualize* numbers takes time to develop, it has nothing to do with categories like unreasonable or illogical. If, on the other hand, a student did not use an estimation strategy and uttered or recorded a guess then the use of *referents* needs to be reintroduced.

- **Relating**: Examples of using correct comparison language as numbers and numerals are compared and ordered.

- **Connecting**: Examples of comments that connect numbers or number names to settings outside the classroom.

- **Confidence** and **willingness to take risks**.

Reporting

The importance of *number sense* makes it essential that information about its development is shared. Oral and written work from students will make it possible to inform parents about possible indicators of the ability to *visualize* numbers and to *think flexibly* about numbers.

There are several advantages to sharing with parents that students have learned to differentiate between guessing and estimating. There are times during conversations when students are asked to make guesses or to report the first thing that comes to mind. All guesses are accepted equally. They are not evaluated in any way. However, when students estimate number they use ten as a *referent* and report their results in the form of, *About ... tens*. Using a *referent* means that students pretend to group the number of objects they are looking at into groups of ten.

Sometimes students are given choices. For example, for twenty four pennies on a plate the choices might be,

About **20 cents**   About **50 cents.**

As the students' number sense develops, the range between the choices is reduced.

To foster *confidence* and *risk taking*, all responses are accepted and they are not evaluated or ranked in any way. Each student's response is used as an indicator of the development of *number sense* for that student. Rating or correcting an estimate does not change the stage of that development.

This type of information provides an opportunity for parents to reinforce what has been learned in the mathematics classroom. It may result in some parents being more careful about the usage of the words guess and estimate. Parents may be more tolerant of responses that differ from their way of thinking, their answers or from what they had in mind. These outcomes would be beneficial to students and to their learning about mathematics.

Numbers and Numerals to Nine Hundred Ninety-Nine

The list of topics, the tasks and the problems that are used to illustrate key ideas for the topics are not suggestive of a teaching sequence. The headings are kept in the same order as for the previous discussions for the sake of convenience.

Types of Activities and Problems

The activities and problems that are described reinforce the importance of the role of a teacher. Appropriate materials need to be selected and prepared, discussions need to be orchestrated and assessment data need to be collected.

Visualizing

Types of requests, tasks and activities that can contribute to fostering the development of *number sense* include:

- Reading three digit numerals in two ways. For example: **345** as '*three hundreds four tens and five ones*' and '*three hundred forty five*'.

- Thinking about the fewest number of base-ten blocks or the fewest number of play money denominations (**$100**, **$10** and **$1**) it would take to show the number for a three-digit numeral. For example: **531** – *Show on your fingers how many blocks or bills you would use to show this number with the fewest number of base-ten blocks. What are the blocks?*

- Explaining how the three digits in a number name with digits that are the same differ. For example, **444**.

- Looking at a three digit numeral and reporting and recording names for numbers that are one hundred, ten and one greater and one hundred, ten and one less that the given number. This request is repeated for numerals with **0** and **9** in the ones and tens place value positions.

- Comparing number names by looking at ten charts: from the **99-Chart** (**0** to **99**); **100-Chart** (**100** to **199**); etc. to the **900-Chart** (**900** to **999**). *How are the charts different? What is the same about the charts? How many different number names are there? How did you figure it out?*

- Looking at a computer printout of dots on sheets that show: one dot; ten dots; one hundred dots and one thousand dots, respectively. *How long do you think it might take to count to one thousand? Explain your answer.* Choices could be presented. For example, *Do you think it would take 2 minutes; 5 minutes or 10 minutes? Explain your thinking.*

(cont'd next page ...)

Flexible Thinking

If one hundred is shown with each of ten students holding up ten fingers, the students can see that **100** or one group of one hundred is the same as ten groups of ten or ten tens.

Representing a three-digit numeral with play money and trading a one hundred dollar bill for ten tens or one ten dollar bill for ten ones shows students other names or an equivalent value for the original number. For example,

    $542   **5($100)** and **4($10)** and **2($1)** or,

             4($100) and **14($10)** and **2($1)** or,

             5($100) and **3($10)** and **12($1)**

A similar procedure is used to have students represent a three-digit numeral in at least two different ways with base-ten blocks.

Rewriting a numeral in expanded form illustrates the meaning of each digit, but does not show a different name for the number.

Estimating

The goal is to provide students with experiences that give them opportunities to use a group of one hundred as a *referent*. These types of tasks do require preparations that involve pre-counting. For example, one hundred names cut from a page of a phone book; one hundred envelopes taken from a box of envelopes; one hundred people from a picture of people in a newspaper. The total number to the nearest hundred of things, objects or people needs to be known. The final preparation step consists of creating two or three choices of possible estimates for the students. Initially the choices should be far apart. This range is narrowed as students gain more experience using one hundred as a *referent*.

For example, one hundred names from a city phone are shown to the students. As these names are held beside one page of the phone book, the students are asked,

> *If there are one hundred names on this part of a page, about how many names do you think are on a whole page? Choose one of these and explain your thinking:*

    About 200 names   **About 400 names**   **About 900 names**

For some examples it may advantageous to present just two choices to the students.

These types of tasks require a collection of pictures of objects or things like cars, trees, people, or animals of known quantities. Once a number to the nearest hundred is known, choices judged appropriate for the students can be prepared. A group of one hundred base-ten blocks (ones) on a paper plate could be used as a *referent*. Students are asked to look at the one hundred blocks, select a choice and be ready to explain their thinking.

These types of activities are required for the ability to *estimate* number to develop. It should be kept in mind that this ability develops gradually as a result of experiencing it in a variety of settings over a period of time.

Students should be invited to try and think about scenarios where, when and why estimation is used in settings outside the classroom.

(cont'd next page ...)

Relating Numbers and Numerals

- The charts, **99-Chart** to **900-Chart**, can be a rich source of activities that involve comparing numerals and thinking about numbers.
  - One number name, two or more adjacent number names in a row or column of different hundred charts are covered. Students are asked to try and think of at least two different ways of explaining how they can figure out what the hidden number names are.

  - Copies of hundred charts are cut up into ten or more pieces. Students are asked to put the pieces together and explain how they knew where the pieces belonged. Initially the cuts are made along rows and columns, but the difficulty level can be increased by making cuts between numerals.

  - Different pieces that look like they were torn from different hundred charts have all but one numeral missing (see page 71). The students are challenged to record the missing number names and be ready to explain in at least two different ways how they know their entries are correct.

  - An *Identify a Mystery Number Name* setting provides opportunities to learn how to ask effective questions and to see the consequences of questions that are asked. A number name from a chart is recorded. The students are told the chart the number name comes from and they are challenged to try and guess it in as few guesses as possible. The rules for asking questions are:

    Questions can only be answered with '*yes*' or '*no.*'

    Questions may not be in the form, '*Is it* ___ ?' (naming a number). For initial attempts a chart showing terms and phrases that could be part of the questions can be provided; i.e., *before, after, between, ones , tens, greater than, less than, close to, odd, even, sum of the digits*.

    Every question has to be different from a question that has already been posed.

  Students are asked to guess how many questions they think it might take to identify the number name.

  The key word or the key words from the questions that are asked and the responses are recorded on the chalkboard. After each question students are to name numbers that could not be the answer and names that could be the answer.

  At the conclusion students are invited to respond and react to:
  - *Did it take more or fewer questions than you had guessed?*
  - *Do you think some questions were better than others?*
  - *Do you think some questions were not effective?*

- Three different digits are presented, i.e., **6**, **7**, **8**. The request is made,
  Think of and print all the different three-digit number names you can make up using each digit once.

(cont'd next page ...)

How can it be determined that all possible number names have been recorded?

Order the number names from least to greatest.

How do you know the order is correct?

What would your rule be for ordering that you could share with a younger student? Compare your rule with the rule another student has written.

How are the rules different? How are they the same?

How many two-digit number names can be printed using these three digits? What are they?

What would you say to someone who suggests that all number names starting with 8 always name numbers that are greater than those starting with 7?

■ *Mystery Chart Tasks:*

• Which Hundred Chart does the secret number name come from?

Here are the choices for the secret number name:

564	482	643	331	228	533
715	657	777	984	866	198

After each hint, the students are asked to explain their thinking for choosing a Chart the number name could be on and naming a Chart or the Charts that can be covered or eliminated.

The hints are:

The number is not greater than 800.
The number is not less than 300.
The number name does not have a 3 in the ones place.
The number name does not have a 7 in the tens place.
The number name does not have a 6 in the hundreds place.
It is not a name for an odd number.
The number name is nearest to 500.
Which Hundred Chart is it on? What is the number name?

• Students are asked to write their own hints for one number name from one of the charts. The list of terms that students can use for their hints include: *greater than; less than; between; ones place; tens place; hundreds place; odd; even;* and *nearest hundred.*

The hints are shared with a classmate. If questions arise about any given hint appropriate editorial changes are made by the author before the hints are presented to the group.

• Students could be challenged to write a *Mystery Chart* type riddle with hints that are not true.

1000 – a Special Numeral: The display of dots: one; ten; one hundred; one thousand, can be used to remind students of the grouping that is done to numbers in our numeration system when the answer to, *How many?* is found.

Ten fingers, or ten ones, are grouped and thought of as 'one student' or 'one group of ten' or as 'one ten and zero ones' – **10**.

Ten groups of ten fingers are put into one group of 'one hundred' and thought of as 'one hundred and zero tens and zero ones' – **100**.

Ten hundreds are put into one group of a thousand and thought of as a 'one thousand' or as 'one thousand and zero hundreds and zero tens and zero ones' – **1000**.

Assessment Suggestions

As students talk and write about numbers and numerals, more information about the development of *number sense* becomes available.

This information can yield further indicators of:

- **Visualizing numbers:** Are students able to describe the fewest number of base ten blocks or bills of the denominations **$100**, **$10** and **$1** it would take to represent a number for a three digit number name?

- **Flexible Thinking about numbers:** Are students able to represent a number for a three digit number name in at least two different ways with base ten blocks or with bills of the denominations of **$100**, **$10** and **$1** (loonies)?

- **Relating numbers and numerals:** Is the language the students use to write their own rules for a mystery number or mystery number name appropriate and correct?

- **Estimating number:** Do the explanations for the strategies the students use to arrive at an estimate make reference to the *referents* they used?

- **Connecting:** Do the comments made during discussions indicate an awareness of who uses these numbers or number names?

- **Confidence and risk taking:** As students talk and write about numbers and number names do they try approaches on their own without being told what to do? As students share what they have written with their classmates, are they willing to explain and defend their thinking?

Reporting

Parents might appreciate knowing that students are able to look at any number name with three digits and that they can identify and tell the fewest number of the denominations **$100, $10** and **$1 (loonies)** it would take to represent that number.

Parents might appreciate being informed that one possible indicator of being able to *think flexibly* is the ability to explain how a number for a name with three digits can be represented in at least two different ways with base ten blocks or with the denominations **$100, $10** and **$1(loonies).**

After students have shared any writing about a mystery number or mystery number name and have edited it according to the suggestions made by classmates, sending this list of hints home to parents will give them an idea about the kind of language students are able to use in their writing as they describe how numbers or number names relate.

Parents would appreciate any comments about observations that can be used as indicators of *self confidence* and *willingness to take risks.*

Fraction Number Sense

The majority of elementary mathematics programs use a region model to introduce students to fractions. At the introductory level fractions are defined as equal parts of regions where equal means the same size and shape. This definition is one reason why some authors use activities that involve 'mirror lines' or 'lines of symmetry as lead-up activities to fractions.

Types of Activities and Problems

Visualizing

- A suitable introductory activity consists of sorting objects, models and diagrams of different figures into one of two groups, those that have or show parts that are of the same size and shape and those that do not.

- Several pieces of paper in the shape of rectangles, squares, hexagons and triangles could be given to the students who are then asked to fold each figure in several different ways. After the pieces are unfolded, they are examined and sorted according to the folds that resulted in equal parts and folds that did not.

- A bulletin board display can show pictures from magazines that were found as a result of an 'equal parts search', diagrams and results of the paper folding that show fractions. The students are invited to make up an appropriate title for the display.

- A 'fraction walk' through and around the school can result in a list that can become part of the display. A numeral is used to indicate how many equal parts there are in each member of the display.

(cont'd next page ...)

- Students are told and shown that the number names they are familiar with are used to name the equal parts. The top numeral tells how many parts are thought of or considered and the bottom number name tells how many equal parts there are.

  Asking students to read the segments between the two numerals as '*out of equal parts*' and later as either '*out of equal parts*' or '*equal number of pieces*' along with the standard name fosters *visualization*. For example, ¼ is read as '*one out of four equal parts* 'and '*one fourth.*'

- *Visualization* is fostered by presenting students with a piece of paper in the shape of a square (rectangle; triangle) and posing problems of the following type,
  > *If the shape you have in front of you is one-fourth (one-third; one-fifth; one sixth) of a figure, what do you think the whole figure could look like? Make sketches and explain your thinking.*

Estimating
The method of reading a fraction in two ways enables students to visualize which of two *benchmarks*, **0** or **1 whole,** a fraction is close to. For example, ¾ or '*three out of four equal parts*' is one part away from one whole.

Flexible Thinking
The ability to visualize fractions enables students to respond to,
> *If ²⁄₄ and ½ are from two figures or objects of the same size, what is different about the pieces and what is the same?*

Relating
Students are asked to assume that each of the following are part of the same region or object: ⁴⁄₁₀ ⁸⁄₁₀ ²⁄₁₀
> *Which is the greatest fraction? Explain your thinking.*
> *Which is the least fraction? Explain your thinking.*

The same assumption is made and the same two questions are posed for:
> ¼   ¹⁄₁₀   ½   ⅛

> *If both ⁹⁄₁₀ and ⅙ come from the same figure or object, which is the greater piece? Explain your thinking.*

Connecting
Students are asked to work with a partner and prepare a list of who uses fractions, when and where? These lists are compared. A composite list is displayed.

Names for fractions – a possible equity issue. Listening to the names of fractions requires fine auditory discrimination. A chart with printed names may be required to show and remind students how certain fraction names differ from the plural form for other similar sounding number names. For example, fours, fourth, and fourths.

The last item on a review that one teacher gave to her students required a response to the orally presented request, *Make a sketch to show five ninths.*

On one student's paper appeared five beautifully drawn stick people that looked like five knights – holding shields and spears. The teacher sketched a happy face and added the comment, *Great!*

Assessment Suggestions

Indicators of *fraction number sense* include:

- **Visualizing and Recognizing:** The ability to recognize and prepare sketches that show fractions and to explain the difference between fractions and non-fractions.

- **Flexible Thinking:** The ability to explain how ½ and ²⁄₄ are the same and how they are different.

- **Estimating:** The ability to tell and explain which of two fractions is closer to zero and which is closer to one whole.

- **Relating:** The ability to identify the greatest and least fraction for a given set of fractions.

- **Connecting:** The ability to tell who uses fractions and where and when.

- **Confidence** and **risk taking**.

Reporting

It could be advantageous to remind parents of the fact that the development of *fraction number sense* is as important a pre-requisite for performing operations with fractions as *number sense* for whole number is for learning the basic facts and for the development of *personal strategies* for computations with the four operations.

Two key indicators of *fraction number sense* are the ability to *visualize* any fraction and to be able to tell whether a fraction is close to zero, close to one-half, or close to one whole. These abilities are fostered by having students look at any fraction and think of it as, *out of equal parts or pieces*. For example, ³⁄₄ is thought of as *three out of four equal parts or pieces*. This awareness by parents can make a valuable contribution whenever fractions may be discussed at home.

For Reflection

What would you say to a teacher who states that teaching to develop *number sense* takes too much effort and time? My students don't really need it. We are just looking at simple arithmetic.

What major points would you include in a presentation to a group of parents about *Number Sense is the Foundation of Numeracy*? How would such a presentation differ for a group of teachers? Identify several indicators of lack of *number sense* and several indicators of presence of *number sense* that you could include in the presentations.

How would you respond to a parent who wants to know why you do not use a number line and the calendar to foster *number sense*?

There exist mathematics programs for students that are based on the key principles of 'direct instruction, repetition and reinforcement.' What questions would you ask of a person who advocates such a program? Why would you ask these questions?

In response to The Daily Special – Math Matters – *Math is never wasted on the Young – Children as young as four months can tell the difference between 1, 2 and 3* [5] one parent included the following statement and request in a letter to the author, *'We are teaching our two-year-old son math by counting, introducing two numbers every two weeks, both written and numerical, in puzzles and game formats and would really appreciate other suggestions for teaching math at this age.'* [6]
What general questions and concerns would you share with this parent?
What possible suggestions would you make?

An introduction to workbooks for a mathematics program [7] includes the following result, *'...children in Grade 2 can learn to perform operations with fractions flawlessly in less than a month,...'* (p.13). Consider the framework of the mathematics curriculum and identify several concerns and questions that could be shared with the author of the statement.

There are three distinct stages of teaching students about the four operations. Initially students are shown how real world situations that involve the *additive, subtractive, multiplicative* and *divisive actions* and the pictorial representations of these actions can be translated into numerical statements. Students will learn about translating numerical statements into sketches that show the appropriate actions and how to create meaningful problem stories for these statements and actions.

At the early stage of learning, the operations are best understood if students associate these with familiar actions from their experiences. This association will foster *conceptual understanding* of the symbols and the mathematical terminology that will be introduced. The introductory activities do not include the recording and reciting of any answers.

After the relation of **is equal to** is introduced, students develop *mental mathematics strategies* that they can use to write equations for the *basic facts* that are true. It will become evident that *number sense* is a necessary and powerful pre-requisite for enhancing this learning process and having students develop *numerical* power.[1]

Number Sense is also essential for enabling students to invent and use *personal strategies* for calculation procedures with numbers and numerals beyond the basic facts.

Understanding Addition

As stories that involve the *additive action* are told and simulated with students, objects or counters, familiar words that describe the action are isolated and listed. Members of the list will include terms like:

> *more came; more were given; some joined; more were put into; received more.*

Students are invited to use these terms to tell action stories from their experiences.

Students are told that it is amazing that the word **plus** can be used to take the place of all of the words that describe the coming together or *adding action*. One grade one teacher described this to her students as almost being like magic – one word can mean all of those things. After the symbol **+** is introduced, students can print summaries for the stories they tell and listen to. Students are able to explain the meaning of each part of the summaries they print: **2 + 3**

- *What does the first numeral tell us?*
- *What does the plus sign tell us?*
- *What does the second numeral tell us?*

(cont'd next page ...)

Types of Activities and Problems

- Writing summaries for dramatizations with students.

- Writing summaries for simulations with counters.

- Making up action stories for given summaries.

- Selecting from several choices the summary that matches a dramatization, simulation or story.

Students need to learn how to identify the *additive action* from diagrams and sketches and to prepare their own sketches that indicate the action. Most student texts start introducing the idea using diagrams with action lines. The diagrams move from the use of familiar things, such as people or animals, to sketches of counters. Other means of showing the intended actions include a hand as part of the diagram, an arrow, or an oval with an attached arrow.

The transfer from simulations with objects to identifying the action in pictures, and from pictures to summaries with numerals must be part of an explicit teaching sequence. It cannot be assumed this important transfer is automatic.[2]

The ability to interpret and draw diagrams and sketches makes new types of activities possible.

- Making up meaningful stories for sketches that include an understanding of:
  - ▸ the numbers,
  - ▸ the order of the numbers: how many to begin with, how many were added;
  - ▸ the *additive action*.

- Preparing sketches that show the numbers, the order of the numbers and the *additive action* for action stories that are presented orally or in written form.

- Selecting from several choices of summaries the one that correctly fits the numbers, order of numbers and the *additive action* included in a sketch. Explaining why the choice is correct and why the other choices are incorrect.

- Selecting from several choices of sketches the one that correctly identifies the numbers, the order of the numbers and the *additive action* in a story problem. Explaining why the choice is correct and why the other choices are incorrect.

Understanding Subtraction

Although different types of situations can be interpreted as requiring subtraction, most programs use the separation of sets to introduce students to the operation. The introductory teaching sequence and the types of activities for understanding subtraction are similar to the introductory sequence for addition.

 ### *Types of Activities and Problems*

The sequence of introductory activities to foster *conceptual understanding* includes the following parts:

- Identifying the *subtractive action* from dramatizations with students, objects and counters and listing phrases that describe the action, i.e., *ran away, lost some, gave away, dropped some, walked away*, etc.

- Substituting the term *'minus'* for the members of the list of action phrases. Introducing the symbol.

- Activities with summaries that have students consider the numbers, the order of the numbers and the *subtractive action*.

- Interpreting diagrams of familiar things and sketches of counters showing the *subtractive action*.

- Preparing sketches for descriptions of actions and providing meaningful descriptions, orally or in written form, for given diagrams and sketches.

- Selecting from several choices of summaries the one that goes with a diagram or sketch and explaining the reasons for the choice and for not selecting the alternatives. If possible, the choices should include one example that might be tempting for someone whose understanding is still developing. The selection of the tempting or incorrect choice provides valuable information about the types of experiences that are required to foster understanding.

  For example: For a rectangular region with four counters and one of the counters enclosed by an oval with an attached arrow indicating the *subtractive action*. The choices are:
  $$3 - 1 \qquad 3 + 1 \qquad 4 - 1.$$

- Students are invited to 'say something about these' summaries:
  $$6 - 3 \qquad 2 - 3 \qquad 5 - 8 \qquad 8 - 5.$$

  During an interview a grade two student shared the observation that, *'Usually you write the bigger numbers first.'* A clarification of *'usually'*, elicited the response that, *'Sometimes the numbers are the same.'* After a pause the student continued by stating, *'and the answers are three, one, three and three.'* This exchange illustrates that a few comments from a student during a conversation can provide valuable diagnostic information.

  If the request 'to say something' is too open-ended for students they are invited to try and make up stories or prepare sketches for the summaries.

- Requesting students to tell or write story problems for summaries can result in some very creative and even amusing results. The word 'story' in such a request results in some students beginning with, '*Once upon a time ...*' without ever returning to anything that was intended unless the request is modified.

The following examples come from files of responses uttered by students during thorough diagnostic interviews.[3]

A student in grade two met the request for a story problem for **5 + 2** with,

> *Once there was a flower who was five years old. She had a brother who was two years old and the mother and father were really old – around 52 and 25. They had a sister who was always getting fed up with the brother.*

When a student in grade three looked at **7 – 3**, the following came to mind,

> *Oh, that's easy. Once there was a seven and his name was John. And there was a three and her name was Karen. They went for a walk to a meadow where there was this big tree. They sat under the tree and sang a song. Can I sing it for you? Love grows under the wild oak tree. Sugar melts like candy. The top of the mountain shines like gold and you kiss your little fellow kinda (sic) handy. After this they went back home and ate a jar of cookies and looked at the clouds.*

After a smile and expressing amazement about these students' ability to see all of that in these simple numerical statements, questions need to be asked. Why are there some students who do not include the order of the numbers or the action in their story problem, even after attempts are made to redirect. It is likely that these students were part of a program where little or no time was devoted to introductory activities and the program focused too quickly on printing and reciting answers. If *conceptual understanding* and transfer to the ability to solve problems is a goal for students, they will have to experience the introductory activities of the type suggested in this section.

Assessment Suggestions

Since the types of records for the *additive action* and *subtractive action* are very similar, the suggestions for these operations are included under the same heading.

As students talk, write and report, indicators of *confidence* and *risk taking* will become available. Assessment information can be noted and recorded about the following ideas and skills:

- Use of appropriate language: The phrases used as the *additive action* or *subtractive action* are described for:
  - Simulations   Diagrams   Summaries

- Matching story problems with given summaries: Do the story problems for summaries demonstrate an understanding of the numbers, the order of the numbers, and the action?

- Matching sketches with story problems: Do the sketches that are prepared for given story problems demonstrate an understanding of the numbers, the order of the numbers, and the action?

- Explaining or illustrating the difference between **3 + 2** and **2 + 3.**

- The reaction to, *Tell me something about **2 – 4.***

Understanding Multiplication

The *equal-groups* interpretation of multiplication makes it easy to make *connections* to previous learning and to experiences outside the classroom. The goals and teaching sequence for introducing students to multiplication are the same as those for the *additive action* and *subtractive action*.

Types of Activities and Problems

- Story problems that require students to move into groups can be dramatized:
  - For one type of problem, students join up in sets that differ in number: one plus three; two plus four; one plus two plus three; etc.
  - For the other type of problem the sets are equal in number: two plus two; three plus three plus three plus three; four plus four plus four; etc.

  The same setting can be created by telling two types of story problems. For one type, counters are placed on students' desks in groups that are equal and for the other type in groups that are not equal in number.

(cont'd next page ...)

For each dramatization or simulation addition summaries are recorded on the chalkboard. The completed list of summaries is examined. To have students focus on the addition of equal groups, the students are asked to identify and describe the summaries they think are in some way the same.

One example is used, i.e., **4 + 4 + 4**, to explain that this summary can be described and summarized as '***three groups of four***' – **3 groups of 4**. The first numeral names how many groups there are and the second how many in each group.

The symbol **x** is introduced. To foster *visualization*, students are asked to read it in two ways, as '*groups of*' and '*times.*' As a further attempt to *connect* multiplication to addition, students could be asked to think of the symbol **x** as the **+** tilted on its side.

- Numerals, the symbol and the two ways of reading can be used to describe the sets of:
    - Wheels on three cars.
    - Wheels on four bicycles.
    - Wheels on five tricycles.
    - Five fingers on four hands.

- The students are shown how an array of dots can be used to illustrate a multiplication summary with the reminder that the first numeral names the number of groups and the second the number in each group.

    For example: **2 x 3**

    When arrays are used the grouping of the members in each set needs to be clearly indicated. Without this clear identification too many possible interpretations exist. It might be possible for four rows of three dots to be interpreted as:

    12 x 1; **1 x 12**; **3 x 4**; **4 x 3**; or, **2 x 6**.

Types of problems can include:
- Writing summaries and drawing arrays for story problems presented orally or in written form.
- Making up meaningful story problems for summaries and arrays.
- Selecting and matching choices of summaries and arrays with story problems.
- Explaining in writing and with sketches how **3 x 5** and **5 x 3** are different. Knowing this difference is important for the *mental mathematics strategies* that are used to learn about the *basic multiplication facts*.

- The following are excerpts from thorough diagnostic interviews about multiplication with four different students from the intermediate grades. The students were requested to make up a story problem for a summary. What are some possible reasons for these responses?

    4 x 3: *My friend had four dollars and I had three dollars. We went to a store and multiplied our money together to see how much we had together.*

    (cont'd next page ...)

7 x 4: *Nancy had seven ice cream cones and Dave had four ice cream cones. No, that's plus. Nancy had seven ice cream cones and Dave had four ice cream cones. Tell me what this is in multiplication.*

2 x 4: *Two people went walking and they found four people and they knew they were walking. They found a stick and made it into an X and they wrote 2 x 4.*

3 x 4: After a request to change this to **4 x 4** was granted by the interviewer, the student said, *We went four by four trucking up the mountain.*

These responses give some indication as to why the students were referred for a diagnostic interview. Several conclusions can be drawn from the responses. The interpretation 'groups of' for the symbol **x** is not known. The students were not able to *visualize* the operation. It is unlikely that these students experienced the introductory activities suggested in this section which are part of the essential content for intervention IEPs for these students.

Assessment Suggestions

As students participate and complete activities, it is possible to collect indicators of *confidence* and *willingness to take risks* and assessment data about ability to:

- Write a summary for a given story problem or sketch.

- Make up a meaningful story problem and a appropriate sketch for a given summary.

- Explain and illustrate the difference between **a x b** and **b x a**.

Understanding Division

Data collected over many years indicate that many students and adults lack *conceptual understanding* of division. What are some possible reasons for this being the case?

Possible factors that are responsible for this lack of understanding include:
- an introduction to the operation that was inadequate as far as fostering *visualization* and *connecting* is concerned,
- use of language not conducive to fostering understanding,
- an early focus on recording and reciting answers,
- rule bound procedural learning.[4]

The Two Types of Division

Students will learn to identify and describe the two distinct types of division actions that are part of their experiences. One type involves *equal sharing*, the other *equal grouping*.

For *equal sharing* settings, a group of objects is shared one at a time. This type of division, partitioning a set one at a time, is referred to as *partitive division*. It can be illustrated with examples like:
- Dealing a deck of cards to players in turn, one at a time.
- Assigning students in turn to several teams, one at a time.
- Placing items of food onto plates in turn, one at a time.

For *equal groupings*, as many groups as possible are removed or subtracted from a sets of objects. Since the group that is removed could be thought of as a 'unit', this is referred to as *measurement division*. Examples that illustrate this type of division include such actions as:
- Assigning groups of five students to make up as many teams of five as possible.
- Filling as many six bottle cartons as possible from a collection of bottles.
- Assigning a group of students to cars, four at a time.

As these types of divisions are dramatized and simulated the students use and hear the phrases '*equal sharing*' or '*shared equally*' for the *partitive division* and '*taking away groups of*' for the *measurement division*.

The dramatizations and simulations are illustrated with sketches that indicate the action that takes place.

For example – *Equal sharing* or *partitive division*:
    There are six students to be assigned to two teams.
    Six dots are drawn to represent the students: ● ● ● ● ● ●
    Two letters are printed to indicate the teams: **A**   **B**
    Arrows are drawn from the first student to team A;
       from the second student to team B;
       from the third student to team A;
       from the fourth student to team B;
       from the fifth student to team A;
       and from the sixth student to team B.

(cont'd next page ...)

For example – *Equal grouping* or *measurement division*:

Six students are arranged into teams of two students.
Six dots are drawn: ● ● ● ● ● ●
An oval with an attached arrow to indicate the taking
away of a group is drawn around each group of two
dots.

Students are told that the summaries for both types of the division actions look the same, **6 ÷ 2**, and are read in the same way, '*Six divided by two*', but depending on the problem type are interpreted differently. For an *equal sharing* problem type it is read as '*six shared equally with two*' and for an *equal grouping* type of problem as '*six take away groups of two.*'

The use of the sign ⌐ can be detrimental, especially if students read the numerals from left to right along with '*goes into*', which according to some authors is equivalent to *mathematical slang.*[5] It is easy to agree with the label *slang*, but the reason for the use of *mathematical* could be questioned since there is not anything mathematical about '*goes into.*'

Why is it necessary for students to know both types of division? Both types are part of students' experiences and mathematics enables students to describe and summarize such events. Both types of division connect to different aspects of mathematics learning.

The *measurement* or *equal grouping* interpretation enables students to develop *mental mathematics strategies* for the *basic division facts* and fosters visualization for certain types of division equations with decimals and fractions. The *partitive* or *equal sharing* interpretation makes it possible for students to develop *personal strategies* for calculation procedures for whole numbers and for division problems for decimals and fractions with divisors that are whole numbers. Not knowing both types of division is likely the main reason why only one of the teachers included in a study was able to make up a technically acceptable, but pedagogically questionable story problem for **1 ¾ ÷ ½.** [6]

Types of Activities and Problems

■ Problem types can include:

- Drawing action sketches for the two types of story problems that are presented orally or in written form.

- Making up story problems for sketches.

- Writing two different types of problems for a summary, i.e., **8 ÷ 2** and illustrating the action for each with a sketch.

- Inviting responses to a summary like **2 ÷ 10**.

- Suggestions for strategies for conversations with students:

A file of diagnostic interviews includes many comments that are indicative of lack of *conceptual understanding* of division. The following are responses to requests for a story problem from different students in the intermediate grades:

20 ÷ 4:   *There were two people called twenty and four.*
Their hobby was dividing.
If you ask them their favourite answer, they say four.

10 ÷ 5:   You have ten apples and you divide them among five children at Halloween.
One child gets six, the others get one each.

14 ÷ 2:   I have never done word problems.
We're doing them next year.
The answer is seven.

The face saving comment from the last student is almost as good as the student who responded to the request for a story problem with, *'My friend was distracting me when the teacher talked about it. I sure hope she will do it again.'*

Attempts during diagnostic interviews to find out whether both types of division are known and which type a subject might prefer can be a challenge. Many students as well as adults want to or do work backwards from an answer and they end up confusing the quotient and the divisor as part of their explanations.

One attempt to prevent subjects from working backwards from an answer is to present a handful or a jar full of counters along with the division sign and a divisor: ⬚ **÷ 3 = __**. The division sign and the divisor are pointed to as the request is made,
If you were to do this to the counters in the jar,
what would you do?

The pointing makes it possible to keep the request free of any mathematical terminology and enables an interviewer to use the language introduced by the subject. The intent is to find out whether a student suggests removing groups of three, the *measurement interpretation,* or removing one at a time and placing them into one of three bunches, the *partitive interpretation.*

A response to a request for a story problem for the equation provides more diagnostic or assessment information, i.e., is the *divisive action* described in the problem consistent with the action used to find the quotient?

After one procedure is suggested by a student, the student is asked whether or not the task can be solved in a different way.

Many students who lack *conceptual understanding* of division will suggest that the equation as presented cannot be solved unless they are allowed to count the objects first.

Assessment Suggestions

Aside from trying to collect indicators of *confidence* and *willingness to take risks*, the main intent of the tasks is to find out whether students are familiar with both types of division and can interpret these concretely, pictorially and symbolically.

Data can be collected about ability to:

- Interpret the two types of sketches that are presented by making up matching story problems.

- Make up two different types of story problems for a given summary. i.e., **8 ÷ 2**, simulate the actions with counters and draw sketches.

- Make appropriate comments about a summary like **2 ÷ 6**. The mathematical slang '*goes into*' and reading division from left to right for the sign ⌐ leads many students to conclude that **6 ÷ 2** '*six divided by two*' is the same as **2 ÷ 6** or '*two into six*' or '*six divided by two.*'

Teaching About the Basic Facts

Teaching about the *basic facts* means that students will learn to:

- use *mental mathematics strategies* so that they can teach the basic facts to themselves.
- use *mental mathematics strategies* to show, in more than one way, that the answers they give and record are correct.
- use *mental mathematics strategies* to reinvent facts that are forgotten.
- get unstuck when faced with not knowing a solution and to do so with a minimum of counting.

The acquisition of *mental mathematics strategies* will result in students who are confident about their knowledge of the *basic facts. Appropriate practice* settings will result in committing as many *basic facts* as possible to memory and contribute to other aspects of mathematics learning as well.

Equality

Before *summaries* are recorded as *equations*, aspects of the *equality relation* need to be introduced. Students need to know that an *equation* is a mathematical sentence written with the symbol =, which is read as '*is equal to.*' The symbol does not mean, '*the answer is*', '*makes*' or '*produces*', but rather expresses an *equivalent relationship*. Reading and interpreting **4 + 3 = 7** as '*four and three is equal to seven*' relates **4** and **3** and **7** and vise versa.

(cont'd next page ...)

Equations are *true* or *false*. This implies that many different responses for equations like **4 + 3 = ■** or **6 = 2 + ■** are possible. The box for these types of *equations* can be thought of as a number sorter. Only one number maintains equality and results in an *equation* that is *true*.

As an activity, one side of an equation is hidden and students are asked to identify number names or different combinations of number names that they think could be hidden to make the equations true. For example: 3 + 4 = ███   5 = ███. Boxes or letters can be used instead of covers as discussions continue about choices of numerals or numerals and symbols that result in equations that are true or false. This type of an activity illustrates what it means to maintain equality and is different from a 'rule' many people recite when they look at equations, 'Whatever you do on one side you must do on the other side.'

Basic Addition Facts

There are one-hundred *basic addition facts*. These facts include all possible single digit combinations, from **0 + 0 = 0** to **9 + 9 = 18**. The importance of *number sense* is illustrated and reinforced by the specific goals that are identified. Without *number sense* the task of memorizing these facts can be quite a challenge and a very boring chore. The absence of *number sense* may result in a program that focuses on endless repetition and timed speed tests.

The learning outcomes for the basic *addition facts* include the ability by students to:
- use the aspects of *number sense*: *recognition, flexible thinking*, and *visualization* to print equations with answers of ten or less.
- generate rules for recording answers for *addition fact* equations involving *zero* and *one* and test these rules.
- use *flexible thinking* about numbers that allows for the *going first to ten and then beyond* strategy for calculating sums.
- use *flexible thinking* about numbers to develop a strategy for calculating sums for *doubles*.
- recognize equations with addends that are *almost doubles* and apply the knowledge about *doubles* to calculate sums.
- recognize equations that have *addends that differ by two* and use *flexible thinking* about numbers and number names to change these addends to *doubles* in order to calculate sums.

Types of Activities and Problems

■ The ability to *recognize* the number shown with fingers on two hands easily transfers to writing *basic addition facts* equations. The movement of the fingers on one hand toward the fingers on the other hand can be recorded as an equation. For most displays of fingers, another different equation can be recorded for the movement in the opposite direction.

For example, for the number seven shown with five and two fingers, respectively, the additive actions will result in: **5 + 2 = 7** and **2 + 5 = 7**.

Flexible thinking about numbers, in this case seven, results in two more equations. Knowing that seven can also be shown with three fingers on one hand and four on the other will yield: **4 + 3 = 7** and **3 + 4 = 7**.

- *Visualization*, or knowing how many fingers cannot be seen or how many fingers need to be added to show the number five, or the number ten, can be translated into *basic addition facts* equations. Students who are looking at two fingers and know that it takes three more to show five and eight more to show ten, can write the *basic addition facts* equations: **2 + 3 = 5** and **2 + 8 = 10**.

- The ability to derive and test *generalizations* is part of *mathematical reasoning*. Students should be given the opportunity to make up their own rules for calculating the answers when one of the two addends is **0** or **1**, respectively.

  What would your rule be for calculating the answers when one
  (zero) is added to a number?
  Compare your rule to the rule written by someone else.
  How are the rules the same? Are they in any way different?
  Do you think your rule would be true for any number?
  Explain your thinking.

This introduction to the *basic addition facts* that is suggested clearly illustrates the importance and power of *number sense*. *Recognition* of numbers, *flexible thinking* about numbers and being able to *visualize* numbers enables students, without much effort or rote practice, to record and teach themselves **sixty-four** of the **one hundred** basic addition facts.

- What *mental mathematics strategies* can students use to teach themselves the remaining **thirty-six** basic addition facts? Knowledge of the answers to the doubles, **a + a = ■**, facilitates the development of *mental mathematics strategies* for these remaining *basic addition facts*. Finger-flash activities help students with *recognition*, *visualization* and recall of the answers for doubles to **5 + 5 = 10**. For the remaining doubles, **6 + 6 = ■; 7 + 7 = ■; 8 + 8 = ■; 9 + 9 = ■**, two possible *mental mathematics strategies* can be used by students.

For one strategy, one addend is renamed to make use of the familiar *how many are needed to show ten* setting to *first go to ten and then beyond*. For example, for **6 + 6 = ■**, one 6 is thought of as **4 and 2**, and the answer is calculated by using the **4** to think of **10** and then of **10 and 2** or **12**.

For a second strategy, two students are selected to use familiar and easy to recognize finger arrangements for the addends. The students are requested to think of an easy way to determine the answer for the double they see displayed. For example, for **7 + 7 = ■** each 7 is shown with five and two fingers, respectively. One easy way to determine and show the answer is to join the two fives and then think of four more, of **10 and 4** or **14**. The students are challenged to respond to,

  How is it possible to use this procedure without having another
  student available or by just showing the number seven?
  What skip counting procedures could be used for the seven
  fingers, or five and two, to arrive at the answer?

It does not take long for many students to suggest to double the fives and then add two twos or four. Some students will require more recognition type activities that will enable them to assign names to arrangements of fingers showing **10 and 2**, **10 and 3**, ..., **10 and 8** without having to count.

The students are requested to try both of the strategies going first to ten and then beyond and skip counting, and to be ready to report which strategy they prefer and why.

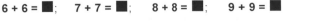

- A list of ten equations that includes five examples that are *almost doubles*, or with addends that have a gap of one between them is presented. After the students have identified and underlined the examples that are *almost doubles*, they are asked to explain how knowing the answers for *doubles* can help with recording the answers for equations that are *almost doubles*.

- A list of ten equations that includes five examples that have a *gap of two* between the two addends is presented. After the students have identified and underlined the examples, two students are requested to show the addends for one of the equations on their fingers, i.e., $7 + 9 = $ ■. To get rid of the gap of two, one finger from the student showing nine fingers is assigned to the arrangement of seven fingers. The result is the display of a familiar double, $8 + 8 = $ ■, and the answer can be recorded. The students are requested to record the matching doubles for the remaining underlined examples.

Appropriate Practice

Appropriate practice settings allow students to use the *mental mathematics strategies* they are acquiring. Settings that place an emphasis on speed are detrimental for reaching this goal, in fact, no mathematics is learned in such settings.

Special instructions or requests are required to foster the development of *number sense* and *operation sense*. Simple instructions like, *Add*, *Solve* or, *Find the Sum* are inadequate. Practice is appropriate when it involves aspects of *reasoning*, *communicating*, *connecting* or *problem solving*.[1]

- Students could look at a sheet of fifty equations and identify the equations that they think are in some way the same by underlining them in the same colour. What strategies would they use to calculate the answers for each of the categories of examples they have identified? The strategies could be explained orally or in writing.

- For five different equations students are asked to explain in writing and with sketches how they would teach someone how to calculate the answers. Students may need to be reminded to think of strategies other than counting.

- Several equations of the type $13 = $ ■ $+ $ ■ are presented. For each example, the students are asked to record as many true equations as they can think of selecting from the digits **0** to **9**. Do they have a way of deciding whether all possible combinations have been thought of?

- Another way to appeal to students' imagination and to get them to think, is to have them try to generate and record, by themselves or with a partner, a list of different types of activities for practice sheets.

- A request to make up a game that uses a practice sheet as a game board gives students a chance to be creative.[7] Testing any game rules that are generated, making revisions, and keeping a written record of all changes and the reasons for the changes involve reflection and further writing.[8] The following is an example of a game designed by a girl in grade three for a practice sheet of fifty four *basic addition facts* in four columns, two columns of fourteen facts and two of thirteen facts. The first item was **8 + 8 =** and the last **3 + 7 =**.

My Rules – Players 3 to 4

- *The idea of my game is to collect as many numbers as you can and to get to the end.*

- *One player starts at the first question (**8 + 8 =**) and adds it. If that player gets it right, he gets it (**16**). Then he counts to **16** any way he wants, across or down. If that question has one of the same numbers as the last question, then it's the next person's turn. If it doesn't have the same number, then continue.*

- *If a player gets it wrong, then it is the other person's turn.*

- *It ends when someone lands on **3 + 7**. Then everybody adds up their numbers, and the person who has the most and the player who lands on **3 + 7** wins.*

- *When it's the next person's turn, they start at where the other person left.*

Creations of this type could lead to a variety of follow-up activities.
For example:

- After the game is played with someone, what rules might be changed and why?
- After the game is played what new rules might be added?
- What rule could be added to ensure that a player who gets an incorrect answer learns not only the correct answer but also how to calculate the answer.
- If the same activity sheet is used and a game that is very different from this game is made up, what would the rules be? What are the major differences between the two games?
- If you were to choose a name for the game, what could be two or three suggestions?
- What general instructions or reminders could you give to players who want to play your game?

Assessment Suggestions

The intent is to collect data about the *mental mathematics strategies* students have at their disposal to get themselves unstuck if an answer is forgotten or unknown or to explain that an answer is correct. The explanations students provide will yield indicators of *confidence, willingness to take risks* as well as *number sense*.

Types of assessment tasks can include:

- Pretend you forgot the answers. Explain how you would figure out the answers for: **9 + 6 = ■**   **8 + 5 = ■**.

- Try to think of two different ways of explaining to someone that the answers for the following equations are correct: **9 + 7 = 16**   **6 + 8 = 14**.

- Describe the different ways you know for calculating the answer for **9 + 8 = ■**.

Basic Subtraction Facts

The *basic subtraction facts* can be thought of as the inverse of the *basic addition facts*. That means that there are one-hundred *basic subtraction facts*: $a - b = c$, where **a** is equal to or less than **18, b** is equal to or less than **9,** and **c** is equal to or less than **9**.

The goals for teaching about the *basic subtraction facts* are illustrated by the responses elicited during a diagnostic interview from a student in grade three . When asked how she knew that the answer **five** she had given for **12 – 7** was correct, she stated, '*I added up from seven.*' Her response to, '*Do you always do it that way?*' was, '*No, for small numbers like one, two three or four* – here she inserted a face-saving comment, '*Pretend I don't know these answers*', twice – '*I kind of count down, but since this is bigger than four I add up.*'

This student's responses illustrate why the *basic addition facts* and *basic subtraction facts* should not be introduced at the same time. Confidence with the *basic addition facts* is a pre-requisite for recording answers for the *basic subtraction facts*. This confidence will enable students to record the answers for every *basic subtraction fact* from a known *basic addition fact*.

Indicators of lack of *number sense* and lack of *confidence* with the basic addition facts surface during diagnostic interviews when students try to solve an equation like **13 – 7 = ■** by starting to count backwards by one on imaginary fingers. When the counting finally ends up on a finger, in this case the seventh finger, students do not know whether that finger, seven, or a finger next to it, six or eight, is the answer for the equation. It would be much more appropriate to count up from seven and keep raising a finger as the count is continued. This makes it easy to recognize the answer.

As part of an activity students can be requested to provide matching addition equations for any basic subtraction fact equation that is provided. The procedure does allow for a guess and test strategy for students who are still developing their *mental mathematics strategies* for the *basic addition facts*.

For example,
>   For **13 – 7 = ■** a student could begin by trying **7 + 3 =; 7 + 4 =**
>   and continue to make adjustments, until arriving at **7 + 6 = 13**

Assessment Suggestions

The main aspect of the data collection includes information about how students connect the *basic subtraction facts* to knowledge of *the basic addition facts*. Questions can also provide information about ability to generalize and to apply these generalizations. As students explain their strategies, indicators about *confidence* and *willingness to take risks* can surface.

- What are the answers? How do you know your answers are correct?

  $17 - 8 = \blacksquare$ $16 - 9 = \blacksquare$

- Pretend you forgot the answer for $13 - 8 = \blacksquare$.
  What would you do to figure out the answer?
  If you were to use your fingers to show how to find the answer,
  how would you do it?

- Make up your own rule for calculating the answers for:
  Subtracting zero from a number.
  Subtracting one from a number.
  Subtracting a number from itself.
  For each rule that is generated, students are invited to resond to,
  Do you think your rule is true for all numbers? Explain your answer.

Basic Multiplication Facts

There are one-hundred *basic multiplication facts*. These facts include all possible single digit combinations, from $0 \times 0 = 0$ to $9 \times 9 = 81$. In the primary grades students examine basic multiplication facts to $5 \times 5 = 25$.

The *number sense* students are developing will allow them to write answers for equations. *Visualizing numbers*, *connecting* to previous learning, *connecting* to experience and *relating* numbers will make it possible for students to begin to develop *mental mathematics strategies* for learning the *basic multiplication facts*.

Types of Activities and Problems

■ Students will be able to print the answers for *two groups of any number* or, $2 \times a = \blacksquare$, by using their knowledge of the doubles from addition. As students record answers for:

$2 \times 2 = \blacksquare$ $2 \times 3 = \blacksquare$ $2 \times 4 = \blacksquare$ $2 \times 5 = \blacksquare$

they are requested to illustrate each equation with a sketch and with a meaningful word problem from their experience.

(cont'd next page ...)

- Students are requested to explain their thinking for;
  If you know **2 x 4 = 8**,
  how can you use this answer to find the answer for **3 x 4 = ■**?

  Use the same strategy to record answers for:
  3 x 2 = ■   **3 x 3 = ■**   **3 x 5 = ■**
  Draw sketches and make up a word problem for each equation.

- Students can use one of two strategies for recording answers for 'four groups of a number' or **4 x a = ■**:

  From the known answer for 'three groups of a number' or
  3 x a = ■, the number for one more group is added.

  Students can also double the answer for **2 x a** to record the answer for **4 x a**.

  If **2 x 4 = 8**, then the eight is doubled to record **4 x 4 = 16**.
  Students are asked to use the two strategies to record answers for:
  4 x 2 = ■   **4 x 3 = ■**   **4 x 4 = ■**   **4 x 5 = ■**
  and explain which strategy they prefer and why.

- The students are asked to record the equation for a word problem that is presented. For example, *How many legs are there on five horses*?

  The students, working alone or in pairs, are challenged to try and think of at least two different ways of finding the answer to the equation **5 x 4 = ■**. The strategies are shared and compared.

  The students are asked to try and find the answers for:
  5 x 2 = ■   **5 x 3 = ■**   **5 x 4 = ■**   **5 x 5 = ■**
  in at least two different ways and to report the reasons for the strategy they prefer.

- Types of problems can include:
  - Matching word problems with equations.
  - Matching sketches with equations.
  - Writing meaningful word problems for equations.
  - Preparing sketches for equations.
  - Writing equations for sketches.
  - Thinking of examples and writing the equations for possible problems that involve *'groups of one.'*
  - Creating a possible problem and writing the equation for *'groups of zero.'*

After students have described a strategy for calculating the answer they are invited to think of another possible strategy.

- Three times a number.
- Four times a number.
- Five times a number.

As students explain their strategies indicators of *number sense*, *confidence* and *willingness to take risks* may be observed and noted.

Basic Division Facts

Basic division facts can be thought of as the inverse of the *basic multiplication facts*. Since division by zero is undefined, there are **90** *basic division facts*; $a \div b = c$ where **a** is less than or equal to **81**, **b** is less than or equal to **9**, but not **0**, and **c** is less than or equal to **9**.

From the introduction to the *divisive actions*, students know both, the *equal sharing* and the *equal grouping* interpretation of division. The latter is best suited for introducing the *basic division facts* since it *connects* to subtraction and multiplication and facilitates the development of *mental mathematics strategies*.

Types of Activities and Problems

A problem is introduced to the students. For example,
Each bicycle frame requires two wheels.
How many bicycles can be constructed from the eight wheels? **8 ÷ 2 = ■**

Students can check to find out if the answer they recorded is correct by multiplying. They can simulate the action of *taking away groups of two* with counters to show that *four groups of two* can be taken away from eight or, **8 ÷ 2 = 4**.

Students know this answer is correct because **4 x 2 = 8**, or *four groups of two* are equal to eight.

This suggested procedure and interpretation is indicative of the fact that *confidence* with *basic multiplication facts* is a pre-requisite for the introduction of the *basic division facts*. The 'equal grouping' interpretation allows students who are still gaining confidence with their basic multiplication facts to use the guess and test strategy for finding quotients.

(cont'd next page ...)

For example,
How many chairs with four legs can be put together from a set of twenty-four legs?
24 ÷ 4 = ■. A student could begin testing with three chairs (**3 x 4**); go on to four chairs (**4 x 4**); then to five chairs (**5 x 4**) before recording the answer six chairs, since **6 x 4 = 24**.

Students should also be given the opportunity to illustrate the solution procedures for the *basic division facts* by sketching action diagrams and simulating the action with counters.

Types of problems can include:
- Matching word problems with equations.
- Matching sketches with equations.
- Writing meaningful word problems for equations and action diagrams.
- Preparing action sketches for equations.
- Writing equations for action sketches.
- Making up and testing a rule for finding the answers to equations that involve *taking away groups* of one.

Assessment Suggestions

Requests are made that can provide data about ability to *visualize*, ability to *connect* and the development of *mental mathematics strategies*. As students explain their strategies indicators of *number sense*, *confidence* and *willingness to take risks* may be noted.

Possible types of requests:
- How do you know the answers are correct?
  12 ÷ 4 = 3   20 ÷ 4 = 5

- Explain how you would calculate and check the answers for:
  15 ÷ 3 = ■   20 ÷ 5 = ■

- Draw a sketch that shows how to calculate the answer for:
  12 ÷ 3 = ■.
  Make up a word problem for the equation.
  Show the action in the sketch with counters. Explain what you are doing.

Reporting

Parents should be informed about the most important learning outcomes for the basic facts. The ability to recall the basic facts is a goal, but more important are the abilities to:
- re-invent a basic fact that is forgotten,
- calculate the answer for a fact that is not known, and
- know at least two ways, other than counting, of showing and explaining that an answer is correct

To reach these goals students develop *mental mathematics strategies* that make it possible for students to teach themselves the basic facts.

To gain insight into what students have learned, requests of the following type could be made:
- After an answer is given, students are asked to tell how they know the answer is correct. Do they know of another way to show and explain that it is correct?

  6 + 8 = 14   13 − 7 = 6
  3 x 4 = 12   12 ÷ 2 = 6

- Ask the students to pretend they have forgotten an answer. How would they figure out the answer?

  9 + 6 = ■   15 − 9 = ■
  4 x 5 = ■   16 ÷ 4 = ■

- Ask how they might teach someone how to calculate an answer that is not known.

  7 + 5 = ■   12 − 5 = ■
  2 x 3 = ■   20 ÷ 5 = ■

Algorithmic Thinking − Addition: *Personal Strategies*

At one time students were taught specific steps or rules to follow as they performed computational procedures. Students were told what to say and record and in what order. The thinking was done for them as they memorized the steps for the standard calculation procedures or algorithms. When asked to explain what they were doing, for many students as well as adults, the recitals almost sound like chants. During diagnostic interviews many students will begin their explanations with comments like,

First you have to line up the numbers.
Then you have to start with the numbers in the ones place.
Write down ....

These students are giving back the words of the teacher. During a conference presentation, Lola May, a well-known mathematics educator, suggested that one reason so many students have difficulties with learning mathematics is that students have to memorize too many of these chants. She claimed that these chants are associated with what amounts to *dance steps*. According to her, when memory fails or gets overloaded, students begin to match '*incorrect dance steps with chants*' or vice versa.

The ability to *visualize* numbers will enable students to develop their own *personal strategies* and recording procedures for computations beyond the basic facts.

(cont'd next page ...)

After a meaningful problem is presented, students can be requested to suggest ways of finding and recording the answer to the problem. The strategies they devise are discussed, compared and a conclusion is reached. For example,

> *Seventeen books were ordered for our classroom library the last time.*
> *Twenty-five more books were ordered.*
> *How many books were ordered?*

To foster *visualization*, two students are requested to show the number seventeen with their fingers and three students can be asked to show the number twenty-five. The class is invited to think of strategies for finding the answer, record their thinking, and be ready to explain and illustrate the thinking with base-ten blocks. As the strategies that were used are compared, students will realize that it is possible to find answers in different ways. However, the different strategies require that the ones and the tens are kept track of in some way.

Types of Activities and Problems

- Students are requested to solve each of two equations in two different ways and to be ready to report which strategy they prefer and why.

   13 + 48 = ■   35 + 45 = ■

- The request is made to create two different equations with two-digit number names that have an answer equal to fifty. Students are asked to explain, orally or in writing, how they selected the two addends. The stipulation that all of the digits in the ones place value positions have to be different or that all of the digits have to differ can be introduced.

   ■ + ■ = 50   ■ + ■ = 50

- Students are asked to try to think of three two-digit number names with different number names in the ones place that when added are equal to fifty. Students are asked to explain the strategy they used for selecting the number names.

   ■ + ■ + ■ = 50

- The request is made to write out in words how to calculate the answer for an equation with base-ten blocks or with students using their fingers and to make up a matching word problem for the equation. A classmate is invited to read the problem and to follow the instructions for calculating the answer. Did any sentences require rewriting? Why or why not?

   29 + 56 = ■

- The request is made to use a calculator and to figure out the answer for an equation in at least two different ways. The students are asked to keep a record of the number names they entered and the order of how these were entered. As classmates report their methods, students are asked to think of how the reported strategies differ and how they are in some way the same.

   29 + 56 = ■

(cont'd next page ...)

- ■ What is wrong? Why do you think it is wrong? Try to make a guess.

   a) For **18 + 35 =** 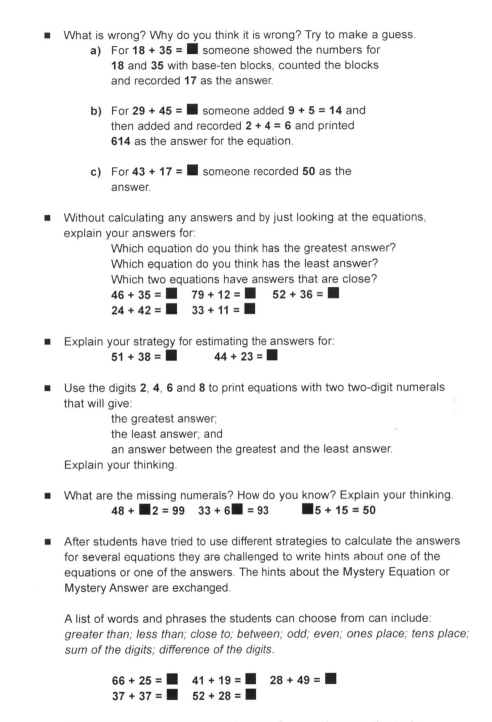 someone showed the numbers for **18** and **35** with base-ten blocks, counted the blocks and recorded **17** as the answer.

   b) For **29 + 45 =** ■ someone added **9 + 5 = 14** and then added and recorded **2 + 4 = 6** and printed **614** as the answer for the equation.

   c) For **43 + 17 =** ■ someone recorded **50** as the answer.

- ■ Without calculating any answers and by just looking at the equations, explain your answers for:

   Which equation do you think has the greatest answer?
   Which equation do you think has the least answer?
   Which two equations have answers that are close?

   46 + 35 = ■   **79 + 12 =** ■   **52 + 36 =** ■
   24 + 42 = ■   **33 + 11 =** ■

- ■ Explain your strategy for estimating the answers for:

   51 + 38 = ■   **44 + 23 =** ■

- ■ Use the digits **2**, **4**, **6** and **8** to print equations with two two-digit numerals that will give:

   the greatest answer;
   the least answer; and
   an answer between the greatest and the least answer.
   Explain your thinking.

- ■ What are the missing numerals? How do you know? Explain your thinking.

   48 + ■**2 = 99**   **33 + 6**■ **= 93**   ■**5 + 15 = 50**

- ■ After students have tried to use different strategies to calculate the answers for several equations they are challenged to write hints about one of the equations or one of the answers. The hints about the Mystery Equation or Mystery Answer are exchanged.

   A list of words and phrases the students can choose from can include: *greater than; less than; close to; between; odd; even; ones place; tens place; sum of the digits; difference of the digits.*

   66 + 25 = ■   **41 + 19 =** ■   **28 + 49 =** ■
   37 + 37 = ■   **52 + 28 =** ■

   Similar introductory activities and types of appropriate practice tasks can be presented for examples that involve adding a three-digit numeral to a two-digit numeral, i.e., **425 + 56 =** ■ and adding two three-digit numerals, i.e., **397 + 576 =** ■.

   Base-ten blocks are used to foster *visualization* and to simulate the action for these types of examples.

Assessment Suggestions

As students supply answers orally or in writing, indicators of *number sense (visualizing, flexible thinking, relating, connecting, estimating)*, *confidence* and *willingness to take risks* may be noted.

Possible types of assessment tasks:

1) 35 + 52 = ■
   a) **Make up a word problem**. Is the problem meaningful? Is reference made to the additive action? Are the numerals used in the correct order?
   b) **Explain your strategy for finding the answer**. As part of the explanation are digits correctly referred to as ones and tens. The explanation could be done orally or in writing.
   c) **Show your strategy with base-ten blocks**.
   d) **Try to think of another strategy to calculate the answer**.

2) 29 + 47 = ■
   a) **Make up a word problem**.
   b) **Explain your strategy for calculating the answer**, orally or in writing.
   c) **Show your strategy with base-ten blocks**.
   d) **Try to think of another strategy to calculate the answer**.

3) **What do you think these students did wrong?**
   25 + 36 = 511   42 + 48 = 80

4) **What are the missing numerals? Explain your thinking.**
   36 + 5■ = 92   14 + ■5 = 99

Algorithmic Thinking – Subtraction: *Personal Strategies*

The students could be introduced to a problem setting like the following:
> *There were fifty-two books on our class library shelves and twenty-seven were taken home one weekend. How many books were left on the shelves?*

With your partner, use base-ten blocks to show the fifty-two books. Try to think of a way of removing or taking away twenty-seven books or blocks. Explain your strategy. These explanations can be illustrated with base-ten blocks on an overhead projector or with six students showing the number fifty-two with their fingers.

(cont'd next page ...)

The strategies that are likely to come up will include:
- Beginning the action by removing two ones. Then removing one ten and returning three ones. Finally removing two tens.
- Beginning the action of taking away by removing two tens or twenty. Then removing another ten and returning three ones.
- Beginning by taking away one ten and returning three ones and then removing two tens.
- Students who have had a lot of experience with *flexible thinking* about numbers and assigning different names to numbers may begin by showing **52**, or **5 tens** and **2 ones** as **4 tens** and **12 ones** and then start by either taking away the tens or the ones.

Accommodating Responses

If every pair of students comes up with the same solution strategy, the students are challenged to try and think of another strategy. After two different strategies are explained by students, another equation is presented, i.e., **61 – 25 = ■**. The students are requested to use both strategies and to be ready to report which they prefer, and why.

The open-ended challenge of trying to think of other possible ways of calculating the answer can be presented. Any new suggestions can then be compared to the strategies that have been tried. Guiding questions may be required to have students think about assigning and recording different names for a number.

Types of Activities and Problems

- Explain two different strategies for taking **26 cents** from **50 cents**. Write an equation. Make up a word problem for your equation. Use sketches of dimes and pennies to illustrate your strategies.

- Try to use addition to calculate the answers for: **48 – 21 = ■** and **89 – 36 = ■**. Explain your thinking.

- Look at the equations. How are they the same? How do they differ?

  $$4 - 2 = 2 \quad 5 - 3 = 2 \quad 6 - 4 = 2 \quad 7 - 5 = 2$$
  $$8 - 6 = 2 \quad 9 - 7 = 2$$

  Try to explain the pattern.
  Why do you think the answers are the same for each equation?

  Add three to each for **52 – 27 = ■** and record the new equation (**55 – 30 = ■**). Why is it easy to calculate the answer for this equation?

  Record the answer for **52 – 27 = ■**.

  If you use the same pattern for **63 – 28 = ■**, which equation will have the same answer? Why is that the case?
  64 – 28 = ■   **65 – 29 = ■**   **65 – 30 = ■**.

  - Use the pattern and write equations that would make it easy to calculate the answers for **81 – 36 = ■** and **70 – 45 = ■**.

(cont'd next page ...)

- Use the digits **2**, **4**, **6** and **8** to print subtraction equations with two two-digit numerals that will give the greatest answer; the least answer; and an answer between the greatest and the least answer.

- Think of at least two different ways to find the answer for **200 − 65 = ■**. How are they different?

- What is wrong? Why do you think a student made these mistakes?
  a) **74 − 36 = 42**   b) **43 − 22 = 25**   c) **50 − 17 = 43**
  After each discussion, the students are invited to respond to, Would you ever make this type of mistake? If the answer is 'no', why not?

- What are the missing number names? How do you know? Explain your thinking.
  a) **39 − 1■ = 16**   b) **■8 − 22 = 66**   c) **6■ − 50 = 10**

- The request is made to write out in words how to calculate the answer for an equation with base-ten blocks or with students using their fingers and make up a matching word problem for the equation. Let a classmate read the problem and follow your instructions for finding the answer. Did any sentences require rewriting? Why or why not?
  56 − 19 = ■

- After students have tried to use different strategies to calculate the answer for several equations they are challenged to write hints about one of the equations or one of the answers. The hints about the Mystery Equation or Mystery Answer are exchanged.

  A list of words and phrases the students can choose from can include: *greater than; less than; close to; between; odd; even; ones place; tens place; sum of the digits; difference of the digits.*
  52 - 38 = ■   **41 - 27 = ■**   **64 - 19 = ■**   **80 - 15 = ■**   **75 - 31 = ■**

- The request is made to use a calculator and to calculate the answer for an equation in at least two different ways. The students are asked to keep a record of the number names they entered and the order of how these were entered. As classmates report their methods, students are asked to think of how the reported strategies differ and how they are in some way the same.
  46 − 28 = ■   **81 − 39 = ■**

Assessment Suggestions

As students supply answers orally or in writing, indicators of *number sense (visualizing, flexible thinking, relating, connecting, estimating)*, *confidence* and *willingness to take risks* may be noted.

Possible types of assessment tasks:

1) 56 – 23 = ■
   a) **Make up a word problem**. Is the problem meaningful? Is reference made to the subtractive action? Are the numerals used in the correct order?
   b) **Explain your strategy for calculating the answer.** As part of the explanation are digits correctly referred to as ones and tens. The explanation could be done orally or in writing.
   c) **Show your strategy with base-ten blocks.**
   d) **Try to think of another strategy to calculate the answer.**

2) 72 – 39 = ■
   a) **Make up a word problem.**
   b) **Explain your strategy for calculating the answer**, orally or in writing.
   c) **Show your strategy with base-ten blocks.**
   d) **Try to think of another strategy to calculate the answer.**

3) **What do you think these students did wrong?**
   $$86 - 28 = 62 \qquad 73 - 52 = 25$$

4) **What are the missing numerals? Explain your thinking.**
   $$6■ - 15 = 54 \qquad 88 - ■6 = 52$$

Reporting

Parents could be reminded that a focus on *number sense* makes it possible for students to develop their own *personal strategies* for computations beyond the basic facts. The *visualization* of numbers and *flexible thinking* about numbers enables students to calculate answers without having to be told how to do it and without having to memorize and recite steps.

Students can explain in their own words, not the teacher's words, how to calculate answers. What they do to find answers makes sense to them. In many cases students can calculate answers in more than one way.

To gain insight into presence of *personal strategies*, students can be asked to explain how to calculate the answers for:
$$35 + 27 = ■ \qquad 40 - 17 = ■$$

For Reflection

What would you say to parents who have their five-year old children practice writing addition and subtraction equations obtained from the Internet?

What possible justifications might teachers present for giving speed tests about the basic facts to their students? What specific learning outcomes might these teachers think they reach with their students? Should students who are unable to print numerals quickly receive some sort of 'handicap' if they find themselves in a speed test setting?

What possible justifications might teachers present for not giving speed tests about the basic facts to their students? What are the greatest possible disadvantages of giving speed tests to students?

What would you say to a grandparent of one of your students in grade two who claims that his grade two grandchild in Europe already knows all of the basic division facts? According to the grandparent schools are much better where he came from in Europe.

How do you answer a parent who wants to know why you have your students look at and draw action sketches for addition, subtraction, multiplication and the two types of division?

What would you say to teachers who tell their students that **2 x 3** and **3 x 2** are the same?

What possible reasons might people give who are in favour of students having immediate recall of the basic facts? What might people say who state that there are more important things than immediate recall?

Chapter 7 – Geometry and Developing Spatial Sense

Many of the ideas, procedures and skills that are part of geometry are a rich source for mathematics problems. One of the major goals of teaching about aspects of geometry is the development of *spatial sense*.

> What is *spatial sense*?
> Why is *spatial sense* important?
> How can the development of *spatial sense* be fostered?

Spatial sense implies *visualization*. This ability to *visualize* includes *spatial reasoning* and *visual imagery*. *Spatial sense* is a necessary part of problem solving.[1] *Spatial sense* is an integral part of *numeracy* and it is not only essential for students' success in mathematics, but is also an important component of other aspects of learning. These aspects include such things as writing letters and numerals; interpreting and preparing tables; making and reading plans and maps; following directions; building models; imagining and describing parts of objects than cannot be seen and *visualizing* objects that are described verbally.[2]

Spatial abilities can be developed and improved. It is possible to create activity settings and orchestrate discussions that are conducive to fostering the development and improvement of *spatial sense*.

Authors who have identified characteristics of settings favourable to fostering the development of *spatial sense* in students have identified such things as: exploring; inventing; discussing in their own words; describing mathematical thinking in discussions with the members of the class; verbalizing their own problem solving strategies; and reflecting on strategies described by others.[3] These characteristics of favourable classroom settings are very similar to those suggested for fostering the development of *number sense* and *operation sense*. The teacher plays a key role in creating and orchestrating these types of favourable classroom settings.

The role of *spatial sense* in problem solving and other aspects of learning suggests that its development should be an integral part of ongoing mathematics teaching and learning. The importance of *spatial sense* suggests that it is advantageous to include activities that foster the development of spatial sense as part of teaching and learning settings at the beginning of a school year.

Learning about geometry in the early grades should be informal and involve explorations, discoveries and problem solving. Whenever possible the main aim of these types of settings should be fostering the development of *spatial sense*. Since it takes time for *spatial sense* to develop, similar types of activities are required in different grades.

Three-Dimensional Figures

There are advantages to begin activities that are intended to contribute to the development of *spatial sense* with a look at *three-dimensional figures* or geometric solids. These figures are referred to as *blocks* from now on.

During the early stages of learning the emphasis should be on activities with *three-dimensional figures*. Students live in a three-dimensional world. Blocks can remind students of objects in their environment. Blocks can be easily manipulated and they can be used to make an easy transition to the exploration of *two-dimensional figures*.

Since the focus is on having students look at *attributes of blocks* and look at similarities and differences between blocks, it is advantageous for a set of blocks to be colourless or for all blocks to be of the same colour. The activities will show that it is necessary for a set of blocks to include duplicates, as well as blocks that are similar as far as *shape* is concerned.

It is possible to foster the development of *spatial sense* without knowing the names of blocks that have a *special shape*. The ability to memorize and recall is not one of the goals of the activities that are presented. Most children are familiar with some names, but many of these names are used incorrectly. There may be some children who may want to know the names of blocks like: *cube*, *sphere*, *cylinder*, *cone*, *pyramid*, and *prism*. There is no harm in learning these names. However, in terms of the specific learning outcomes for three-dimensional figures related to the development of *spatial sense* there are no advantages.

Introductory Settings

The thinking strategies of *sorting*, *matching* and *ordering* play an important role in the development of the ability to *visualize*.

Types of Activities and Problems

- **Connecting – It Depends**
  Students sit or stand around a collection of blocks spread out on the floor or on a table. The request is made to select one block. As the students hold the block in one hand in front of them at eye-level, they are asked,
  What does this block remind you of?

  The same question is repeated while looking at the same block on the floor, while standing over the block, and while holding it with several fingers straight above the head. The students are invited to respond to,
  Why can the same block remind us of different things?

  As the task is repeated with a different block, students are led to the conclusion that includes the idea that, *It depends how a block is looked at.*

■ Sorting – Look Again

Students face a collection of blocks. For a task that involves visual sorting, one block is selected. The students are asked, in turn, to point at one of the blocks in the collection that is:

- Exactly like the block that is held up.
- Like the block, but is bigger. Explain your thinking.
- Like the block, but is smaller. Explain your thinking.
- A little different from the block. How is it different?
- Very different from the block. How is it different?

After each request one student is asked to pick up the block that is pointed to and hold it up. If more than one block is pointed to, these could be held up by different students.

■ Pictorial Recognition

The ability to connect 3-D figures to matching pictorial representations of these figures is an important aspect of the ability to *visualize*. As students face a collection of blocks, a photo or a drawing without any dotted lines to indicate hidden parts is shown on the chalkboard or with an overhead projector. The students are asked to point to a block that they think is shown.

One student is asked to select the block, stand beside the photo or drawing and try to hold the block as it is shown in the photo or drawing. If several copies of the same block are part of the collection, several students are given the opportunity to stand beside the photo or drawing. A student is requested to point on the block to the parts that are shown in the drawing or the parts that can be seen. Then the request is made to point to the part or parts that cannot be seen in the drawing. The procedure is repeated with a different block.

■ Points of View

Out of the students' view, six or seven blocks are used for a simple construction on a platform like an atlas or a book. Several smaller blocks are part of the building that cannot be seen by the students. The students are presented with a frontal view of the construction and are invited to respond to,

> *Show on your fingers how many blocks of this building you can see.*
> *Show how many blocks you think were used for the building.*

If the response for both of the questions is the same, redirection is required. Students need to be led to realize that unless other view-points are available, the second question cannot be answered.

Parts of 3-D Figures: Faces

Assigning a name to parts of a block presents opportunities to make new discoveries, draw conclusions and further develop the ability to *visualize*.

As part of an introduction, every student selects a block. The request is made to feel or trace the big parts of the block with their hand and to try and think of how all of these big parts, which are called *faces,* are the same.

After it is concluded that all *faces* are smooth, students are asked to look at the *faces* of the block and at the *faces* of other blocks and describe how *faces* of blocks can be different. Some guidance and redirection may be required to elicit that *faces* can be round or flat; large or small; and can have different shapes.

Types of Activities and Problems

- **Sorting**
Students are requested to select a block and count the faces on the block.
> *How do you know you have counted all of the faces?*
> *How do you know one face was not counted twice?*

This activity took place in one classroom where eight pieces of paper with the labels:
> **0 faces**; **1 face**; **2 faces**; **3 faces**; **4 faces**; **5 faces**; **6 faces**;
> **More than 6 faces**

were placed on the floor.[4] Students were asked to keep on walking by a collection of blocks, select one block and place it onto the appropriate piece of paper. After all or most of the blocks from the collection were placed onto the appropriate sheets of paper, the students were invited to respond to the open-ended question,
> *What do you notice about the blocks on the pieces of paper?*

The responses by different students included the following observations:
> *'Some pieces have many blocks.'*
> *'Some pieces of paper have few blocks.'*
> *'Two pieces of paper do not have any blocks.'* (**0 faces** and **4 faces**)
> *'The piece of paper with **6 faces** has the most blocks.'*

The students were also asked to respond to,
> *How are the blocks on the different pieces of paper different?*
> *How are the blocks on the same sheet of paper the same and*
> *how are they different?*
The answer to the second question led to the discovery that blocks that are the same as far as the number of faces is concerned can look different.

The students in this classroom were asked,
> *Do you think it is possible to have a block with zero faces?*
All of the students agreed that this was not possible.

Since the sheet of paper labelled **4 faces** did not have a block on it, the students were asked whether they thought it was possible to have a block with four faces. There was agreement that this is possible. Since that was the case, the students were challenged to try and find a block with four faces.

One student made the comment that at first he thought it was not possible to have a block with just one face, but then he remembered that faces can be round.

The excerpts from this lesson reinforce the importance of the role of the teacher who not only plans appropriate activities and creates appropriate settings, but asks high order thinking questions that focus on the development of important ideas and skills.

- **Hidden Faces**

    A photo or a drawing without dotted lines to indicate hidden parts is shown on the chalkboard or with an overhead projector. The students are asked to use their fingers on one hand to indicate the number of faces on the drawing of the block that they can see. Then the request is made to use fingers on the other hand to show how many faces they think they cannot see. If the number shown is the same for every student, one student could be given the opportunity to explain what was imagined or how the number was arrived at.

Accommodating Responses

Care needs to taken if there is a disagreement about the answers about hidden parts. Such a disagreement can easily be spotted by looking at the numbers displayed by the students. If the answers for the total number of faces on the block differ, several reasons for that being the case may exist:

- A simple counting error could have been made.
- Different levels of ability to *visualize* can exist. It is a rather difficult task to keep track of faces to be counted on a block and it is even more difficult if it has to be done mentally.
- Some students may not be ready to *visualize* the hidden parts.
- There may be a student who *visualizes* something other than a regular block from the collection of blocks.

- **Different Possible Answers**

    One block is shown to the students. The block is held in a stationary position a little above the students' eye level. The students are requested to show with fingers on their right hands the answer to,

    How many faces of the block can you see?

    Since the students in a classroom get different views of the block, the answers should be different, which leads to the questions,

    Why do some students show a different number?
    What is the reason?

    The students are requested to show with fingers on their left hand the response to,

    How many faces do you think you cannot see?

    A collection may include blocks that require students to use both of their hands to show their responses to, *How many faces do you think you cannot see?*

Different blocks are held in different ways, and the same questions about faces that can be seen and cannot be seen are posed. The final question is, *Why are the answers different from what they were before?*

Inspection of the fingers that are displayed will tell whether or not every student is visualizing a block in the same way. If the responses are not the same the options listed under the previous **Accommodating Responses** need to be considered.

For example,
A cone can be held in a way that will elicit the responses from students of, *'one'* for the face that can be seen and, *'one'* for the face that cannot be seen. Some collections of blocks contain cones with a piece sliced off. If that part is hidden from the view of the students, there is a different answer to the question,
How many faces do you think you cannot see?
Now the answer to the question is, *Two faces could not be seen.*

This example illustrates several ideas. First, it is important to include *'do you think'* in the question. This inclusion gives students the opportunity to save face because this enables them to explain that they were thinking of a different block. Secondly, students need to be led to conclude that, *It is not possible to predict with certainty what hidden parts may look like.*

Certainty requires more view-points. It is important for students to learn and conclude that there are times when objects or characteristics of objects are quite different from what they appear to be or different from what students think they should be. This latter notion is an important requisite for teaching students about measurement.

- **What Might it Be?**
  A block is wrapped up in a piece of paper.
  Only one face is visible and is shown to the students.

  For example, for a square-based pyramid the square is shown to the students. The same setting can be created by placing a square-based pyramid behind a cardboard screen onto an overhead projector and then turning it on. The initial question could be,
  If this is a face of one of the blocks, who knows which block it is?
  Which block from our collection do you think it is? Why do you think so?

(cont'd next page ...)

During early stages of *visualization* and development of *spatial sense* all students are likely to agree that the block has to be a cube. It will take time before some young students realize that many answers are possible. If this is the case for a group of students, redirection is required and the questions posed need to be revised to,

> *Which of the blocks from our collection could this be?*
> *What are some other possible answers?*
> *What should be done to the block on the overhead projector or behind the piece of paper in order to get more information about the block?*

Visualization can be fostered for students if they are standing over the collection of blocks and look at and select all of the different types of blocks that do have a square face.

■ **Feel the Faces**

Students are assigned to a partner. Turns are taken. One student places a block into the other student's hands behind the back and makes requests to find a block that is:

- Exactly like the block.
- A little different from the block.
- Very different from the block.

For each task a request for an answer to the following question is made,

> *How did the faces help with selecting a block?*

■ **Does Not Belong**

Four students select four different blocks and stand side by side in front of the group displaying their choices. The members of the group are invited to,

> *Try and think of all of the things these four blocks have in common.*

Some guidance and redirection may be required to elicit comments about the facts that:

- they are all blocks;
- they are made out of wood, or plastic;
- they all have faces;
- they all can remind us of something from our experience.

The following request is made,

> *Think of faces and identify one of the four blocks that you think is different or does not belong. How is it different?*

Identifying one block as being different, means that the remaining three blocks belong to the same category of blocks. *Flexible thinking* is fostered if the students are challenged to identify another block that could be identified and described as being different.

As the task is repeated with different choices of blocks, it might be possible for the students to identify a third member of the group as being different.

(cont'd next page ...)

Matching Faces

An activity that can involve *sorting*, *approximation* and the strategy of *guess and test* requires the tracing of the different types of faces of a block onto a sheet of paper. The tracing procedure is repeated for different blocks.

The students work in small groups or with a partner. The students are told that they can pretend that the different faces on a piece of paper have been traced from a block. The challenge is to,

Try and identify the block that has these faces and then determine how many of each type there are.

The request is made that students talk to each other as they examine the collection of blocks and try to decide which block might match the faces on the piece of paper. The discussions that are part or can become part of this suggested setting are free of mathematical terminology; that is, names are not assigned or used for the two-dimensional figures.

Riddles about Faces

As the students look at a collection of blocks, comments about faces are made and the students are asked to respond by pointing to or selecting a block that they think meets the conditions of the comments.
For example,

The block has two different types or kinds of faces.
Which block could it not be and why?
Is there another block that is excluded?
Which block could it be and why?
Is there another block that could be the one?

Some of the questions could be repeated again. Then a second hint is provided,

One type of face is round and the other type is flat?
Which block or blocks could it not be and why?
Which block or blocks could it be and why?

The next hint is presented,

The block has three faces.
Which block do you think it is and why?
Do you think it could be another block? Why or why not?

After responding to several riddles of this type, the students could work with a partner and try to write their own riddles about one block and share it with their classmates.

Following Directions

Pairs of students, in turn, are asked to respond to directions of erecting a building with blocks. For example,

The first block has six faces of two different types.
The second block has six faces of two different types.
The next block has six faces and all of them are the same.
The last block has two types of faces; one is flat and the other round.

(cont'd next page ...)

After several pairs of students have had an opportunity to respond, the buildings they created are displayed.
What is the same about all of these buildings? Why?
What is different about the buildings? Why?

Guidance and redirection may be required to have students realize why these constructions are or can differ in size and why there is more than one correct response for the last condition given.

- **Writing Directions**
  After a few experiences of following directions, students working in pairs can be given an opportunity to write their own directions for a building that can be attempted by classmates.

  Working in pairs, students construct a building from three or four blocks. To enable classmates to try and replicate such a building, written instructions are created that identify the number and types of faces on each block and the order in which the blocks were used.

Spatial Sense Beyond the Mathematics Classroom

Visual thinking or *imagery* can be and is part of many other activities. During classes in the gymnasium, students can face each other and try to copy or mirror the moves made by a partner. The challenge of trying to do the opposite can be presented.

The scenario of copying the action, or doing the opposite, can be introduced while standing behind a partner.

When certain pictures are drawn during Art, students could be requested to try and sketch a different view-point of whatever they were drawing.

The students in a grade one classroom had drawn a side-view of a car that showed two wheels. This seemed to be an ideal opportunity for a teachable moment. One student was approached and the request was made,
Show with fingers how many wheels there are on a car.
After the student had shown four fingers, he was asked about the number of wheels on the car he had drawn. It was more than a little surprising when four fingers were shown again. The student flipped his piece of paper. There were two more wheels on the other side. It was fascinating to note that he was the only student in the class who had done this.

Parts of 3-D Figures: Edges

As students slide a finger along the part of a block where two *faces* meet, they are asked to suggest a name for it. The label *edge* is adopted as the term for this part. Suggestions are elicited for,

> *How are all of the edges the same?*
> *How can edges be different?*

Guidance may be required to help students conclude that these sharp *edges* can be straight, round, long or short.

Types of Activities and Problems

- **Sorting**

  Inspection can be used to sort several blocks into one of two categories, many *edges* and few *edges*.

  As students stand around the collection of blocks they are asked to identify by pointing and selecting blocks that have zero or no *edges*, one *edge*, and two *edges*.
  > *What is the same about the blocks that have few edges?*

  Students are requested to identify blocks that have six *edges*, eight *edges* and twelve *edges*:
  > *What is the same about these blocks?*
  > *How can blocks with the same number of edges look different?*

- **Hidden Faces and Edges**

  A drawing of a block with twelve edges that uses dotted segments to indicate the hidden edges or the edges that cannot be seen is shown to the students. They are invited to suggest why dotted edges are part of the drawing.

  When one student in grade two was asked what the dotted edges in the drawings of his textbook meant he responded with,
  > *'To make an optical illusion.'*

  It may be tempting to dismiss this comment, but the student may not have realized that he had a valid point. For some diagrams of this type it is possible to make the intended hidden part appear as the front part of a figure. After blinking, it can reappear as part of the back of the figure.

  Photos or a drawings without dotted segments to indicate hidden parts are shown on the chalkboard or with an overhead projector. A chart with the following headings is provided for each sketch of a block:

	I Can See	I Cannot See
Faces		
Edges		

(cont'd next page ...)

The students are requested to enter the answers to,

How many faces and how many edges can you see?
How many faces and edges do you think you cannot see?
What are some possible ways of checking to find out whether
your answers are correct?

Since the students are requested to speculate about hidden parts the responses they enter in the chart may differ. If that is the case, the suggestions made for **Accommodating Responses – Hidden Faces** (page 115) will have to be considered.

- **Different Answers**
One block is shown to the students. The block is held in a stationary position a little above the students' eye level. The students are requested to use the chart to indicate how many *faces* and *edges* they can see and how many *faces* and *edges* they think they cannot see.

	I Can See	I Cannot See
Faces		
Edges		

Why are the answers for different students in the classroom different?
How is it possible to find out that entries are correct?

The same block is held up but is rotated a little. The requests for entries in the chart are repeated.

If your entries are different from before, why is that the case?
How do you know your entries are correct?

- **Does Not Belong**
Four students select four different blocks and stand side by side in front of the group displaying their choices of blocks. The members of the group are requested to,

Think of faces and edges and identify one block that you think
is different from the other three blocks. How is it different?

To foster *flexible thinking*, the students are asked to try to identify another block that can be described as being different in some way. This procedure is repeated until the students are unable to provide a response for another member of the collection.

- **Two Differences Line-Up**
One student selects a block and stands at the front of the classroom. A classmate selects a block and is permitted to stand next to the student if the block is described as being different in two ways and reference is made to *faces* or *edges*. The next student selects a block and tries to join the line-up by telling how the selected block is different in two ways. The goal is to have everyone join the line-up and have everyone say something about *faces* or *edges*.

The students can be presented with the challenge,

Let us try to get everyone into a three differences line-up. Every block
has to be different in three ways from the last block in the line-up.

Parts of 3-D Figures: Corners

As students trace the *edges* of a block they have selected, they are asked to think about the part where two or three edges meet. This 'pointy' part can be called a **corner**.

Types of Activities and Problems

- **Sorting**

  Inspection can be used to sort several blocks into one of two categories, many *corners* and few *corners*. Students are asked to try and find blocks that have zero or no *corners*, one *corner*, four *corners*, five *corners* and eight *corners*.

  After the students have found and sorted blocks according to the same number of corners, each set in turn is examined and the students are invited to respond to,

  > *What is the same about the blocks in this set?*
  > *How are the blocks in each set different?*

  Guidance may be required for some students to lead them to conclude that blocks that have common characteristics can look different.

- **Hidden Faces, Edges and Corners**

  A photo or a drawing, without dotted lines to indicate hidden parts, is shown on the chalkboard or with an overhead projector. A chart with the following headings is provided for each sketch of a block.

	I Can See	I Cannot See
Faces		
Edges		
Corners		

  For each pictorial representation the students are requested to enter how many *faces,* how many *edges* and how many *corners* they can see and then how many *faces, edges* and *corners* they think they cannot see. The students are asked,

  > *What are some possible ways of checking to find out*
  > *whether the entries are correct?*

(cont'd next page ...)

- **Different Answers**

  One block is shown to the students. The block is held in a stationary position a little above the students' eye level. The students are requested to use the chart to indicate how many *faces*, *edges* and *corners* they can see and how many *faces*, *edges* and *corners* they think they cannot see.

	I Can See	I Cannot See
Faces		
Edges		
Corners		

  Why are the answers for different students in the classroom different?
  How do you know your entries are correct?

  The same block is held up but is rotated a little. The requests for entries in the chart are repeated.
  If your entries are different, why is that the case?
  How do you know your entries are correct?

  Since the students are requested again to speculate about hidden parts the responses they enter in the chart may differ. If that is the case, the suggestions made for **Accommodating Responses – Hidden Faces** (page 115) will have to be considered.

- **Does Not Belong**

  Four students select four different blocks and stand side by side in front of the group displaying their choices of blocks. The members of the group are requested to respond to,
  Think of faces, edges and corners and identify one block that you think is different from the other three blocks. How is it different?

  To foster *flexible thinking*, the students are asked to identify another block that can be described as different. This procedure is repeated until the students are unable to provide a response for another member of the collection.

- **Two or More Differences Line-Up**

  One student selects a block and stands at the front of the classroom. A classmate selects a block and is permitted to stand next to the student if the block is described as being different in two ways and reference is made to *faces, edges* or *corners*. The next student selects a block and tries to join the line-up by telling how the selected block is different in two ways. The goal is to have everyone join the line-up and have everyone say something about *faces, edges* or *corners*.

  The students can be presented with the challenge,
  Let us try to get everyone into a three difference line-up where every block is different in at least three ways from the last block.

The topological notions related to simple closed curves like inside and outside can be used to introduce students to special closed curves or *two-dimensional figures*. Blocks are well suited for an introductory setting. The result of tracing a face of a block onto a piece of paper or onto a chalkboard shows students that the tracing separates the area of the partial plane or the chalkboard into three distinct parts:

- the tracing or the two-dimensional figure with a special shape,
- the area inside the tracing or shape, and
- the area outside the tracing.

One of the goals at the introductory stage is for students to conclude how *two-dimensional figures* with the same number of sides and angles can differ. Students will also learn that the number of sides or corners of *two-dimensional figures* are used to name figures: *triangles, pentagons, hexagons and octagons.*

Rather than describing the same possible introductory setting for each special shape, a possible introduction to the *rectangle* is described in detail. Similar strategies and activities can be used for the other special shapes.

Rectangles

As students observe, the face of a rectangular prism is traced onto a piece of paper or onto the chalkboard.

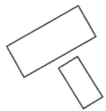

Questions of the following type are posed:

Show with your fingers how many sides the tracing has?
What is the same about the four sides?
Show with your fingers how many corners the tracing has?
What is the same about the four corners?

Students need to be guided to conclude that the four corners all look and are the same, they are all L-shaped, which could be simulated with the left hand, and they are called *square corners*.

Students are told that any figure that has four straight sides and four corners that are the same are called *rectangles*. Any tracing of a rectangular face results in three distinct parts:

- the *rectangle*,
- the inside of the *rectangle*, and
- the outside of the *rectangle*.

There exist references that ask students to *colour a rectangle (triangle, circle)*. According to the correct definition, that is not an easy task. If it is not the intent to have students trace the sides of the figures, the request should be terms of the inside of the figures, i.e., *colour the inside of the rectangle (triangle, circle)*, a distinct difference.

(cont'd next page ...)

The request is made to identify blocks that have rectangular faces that differ from the one that has been traced. The selections made by students are traced. If a cube is not chosen, it could be a choice made by the teacher. As the students look at the sample of tracings, they are asked,

> *What is the same for all of these rectangles?*
> *How are the rectangles different?*

Guidance may be required to have students focus on the special *rectangle* with sides that are the same length. Students are told that these special rectangles are called *squares*.

> *How can special rectangles that are called squares be different?*

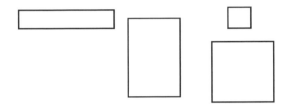

Triangles and Circles

The introduction to *triangles* and *circles* can be similar to the one used for *rectangles*. Young students will have to be led to conclude that any two-dimensional figure with three straight sides and three corners is a *triangle*. It does not make sense to talk about an upside-down *triangle*, as many do, and no matter how short the base and how long the other two sides, it is still a *triangle*. The examples and discussion about how *triangles* can be different is important for reaching a generalization about triangle.

During a discussion about possible differences for *circles*, a comparison can be made to the discussion about the special *rectangles* called *squares*.

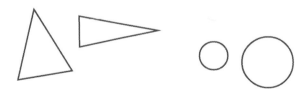

Pentagons, Hexagons and Octagons

If students are introduced to tracings around faces of blocks with five, six and eight sides of the same length and their respective names, a chart is required. Such a chart should remind students that the Greek words beginning with 'penta', 'hexa' and 'octa' make reference to five, six and eight, respectively. The labels *pentagon, hexagon* and *octagon* tell how many sides a figure has.

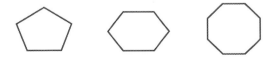

(cont'd next page ...)

Types of Activities and Problems

- **Shape Hunt**

  The goal is to make a list of objects from around the school and outside the school that have the shape of: *triangles; rectangles; circles; pentagons; hexagons and octagons*. Lists are combined.

  > *Which shape is the most common? Why might that be the case?*
  > *Which shape is the least common? Why might that be the case?*

- **Preparing Designs**

  The request is made to design an object or a creature using only one special shape.

  > *How many shapes were used?*
  > *Was this easy or difficult? Explain your answer.*

- **Preparing Patterns**

  The request is made to use variations of one shape and to create two different patterns. If one or two members of a pattern are hidden, can someone tell what they are by looking at the remainder of the pattern?

- **Preparing Designs with a Pattern**

  The students are challenged to use a pattern to create a design for a piece of wall paper, a part of a floor, or a flag for an imaginary country. If parts are hidden, which parts can be identified and which parts cannot? Why is that the case?

- **Models of Special Shapes**

  Building models can foster *visualization* and reinforce important ideas. Students are given pieces of drinking straws of different lengths and pieces of pipe cleaners or Plasticine and are requested to construct as many different *rectangles* or *triangles* as they can. Upon completion, questions of the following type are posed,

  > *How are all of the rectangles (triangles) the same?*
  > *How are the rectangles (triangles) different?*

  The question, *How many sides and how many corners does each rectangle (triangle) have?* can be used to introduce a challenge to investigate the relationship between edges and corners.

  > *Do you think every different shape you construct or sketch*
  > *will have the same number of sides and corners?*
  > *Try a few different shapes. What did you find out?*

Assessment Suggestions

In the early grades the goal is not to have students memorize terms and definitions and to assess students' ability to recall what has been memorized. The main goal of the assessment should be to probe the ability to *visualize*. During activities and as students respond to questions, indicators of *confidence* and *willingness to take risks* may also be noted.

Assessment tasks can provide answers to questions of the following type:

- Are students able to look at a drawing or photograph of a block, select the block that is shown, hold it in the way it is shown and point to the hidden part or parts?

- Are students able to look at a drawing or a photograph of a block and tell how many *faces*, *edges* and *corners* are visible and explain how many *faces*, *edges* and *corners* they think may be hidden? Are they able to explain why 'may be hidden' is included in the question?

- Are students able to match the different types of *faces* of a block that have been traced onto a piece of paper with the appropriate block and tell how many of each type of *face* are on the block?

- Are students able to tell how *faces* and *edges* on blocks are the same and how they can differ?

- Are students able to tell how all *rectangles* (*triangles*; *circles*) are the same and how they can differ?

Reporting

There may be parents who believe that memorizing names for 2-D and 3-D figures is one of the most important learning outcomes of examining aspects of geometry.

During a presentation to parents at the time of writing this chapter, one parent made the comment that she was in the process of memorizing many names. Her reason for doing this was that she wanted to help her daughter who was in grade three and had to learn all of these names. She had not heard of *spatial sense*.

Parents need to be told, in a newsletter or during a meeting, what *spatial sense* is and why fostering its development is an important goal of teaching students about aspects of geometry. Parents can also be informed about possible activities and questioning strategies that they could make part of their conversations. This type of interaction could contribute to fostering the development of *spatial sense* and the ability to visualize.

For example parents could get students to speculate about things that they cannot see or things that are hidden,

- *What do you think the back of the house or barn looks like?*
- *What do you think is behind the fence?*
- *What do you think is on the other side of the tree?*
- *Feel the object in the bag. What do you think it could be?*
- *I am going to put something into your hand behind your back. What do you think it is? What made you think of that?*
- *Look at this drawing of a house (car, train, boat, tractor, etc.). Try to tell me something about the part that is not shown.*

Parents can be informed of indicators of *spatial sense* that were observed during the teaching.

For Reflection

How would you respond to a teacher who states, '*I don't think geometry is really that important; I always leave it to the end of the year and only teach it if I have enough time.*'

How would you define *spatial sense* to a parent or friend and what indicator of *spatial sense* could you use to illustrate your definition?

Explain how you could use an activity or a problem that deals with an aspect of *spatial sense* to build *confidence* and *use of imagination*.

Why do you think many adults have a negative attitude toward geometry?

During a conversation, a student in grade one was asked to try to make a copy of a building of blocks that was standing on a table. After several blocks had been put in place, the student turned away and began to play with other blocks and a car. When asked whether or not he thought that the task that was given to him was too difficult, he responded with, '*No, but I like doing this much better.*' What would you do? What would you say to the student? Why?

Chapter 8 – Measurement

Measurement and Measurement Sense

Counting gives the answer to the question how many by assigning number names to sets of discrete objects. Counting is to discrete quantities what measuring is to continuous quantities. Measuring gives the answer to,
How much according to:
- *How long?*
- *How much does it hold?*
- *How heavy?*
- *How long does it take?*

The process of measuring is very similar for any continuous quantity – a unit is chosen and the characteristic being measured is compared to and reported in terms of that unit.

The majority of young students may see no need to learn the rather contrived skills and ideas that are part of measurement. It is more than likely that measurement is not part of their out-of-the classroom experiences. As far as these students are concerned, they trust their eyes. Things are as they 'see' them and the use of the descriptors *big*, *bigger* and *not as big* and *small*, *smaller* and *smallest* is sufficient for any settings or tasks that require making comparisons or ordering.

It is advantageous for teachers of young students to base their teaching about measurement on the assumption that time will be needed to get these students to see a need for learning measurement skills and ideas, learn the measurement skills and use appropriate descriptors.[1]

As is the case for teaching about geometry and the development of *spatial sense*, learning about measurement in the early grades should be informal and involve explorations, discoveries and problem solving. The instructional settings that are favourable for the development of *measurement sense* are similar to those described for the development of *number sense* and *spatial sense*. Students need to be active, communicate in oral as well as in written form and have opportunities to compare strategies as well as results.

The ability to measure involves the use of such skills as iteration or repeated application of a unit and transitive thinking for comparisons. *Measurement sense* includes:
- knowing the appropriate units for a task.
- knowing when to measure and when to estimate.
- knowledge of several estimation strategies.[2]

The topics for the primary grades in the curriculum include measurement of length, capacity (volume), mass and time. Perimeter is included as a separate topic, but it is part of measurement of length.

(cont'd next page ...)

Since young students rely very heavily on perception, the world to them is as they see it. They see no need for measurement of any kind. These students need to learn that sometimes things are not as they appear. Their eyes could deceive them and they need to do something to find out what is true. For example, it could be that two differently shaped pieces of cake contain the same amount of cake even if one appears to be more or, two identical looking parcels could be quite different as far as content and weight are concerned.

The basic measurement ideas and skills are similar for all topics. Since that is the case, an instructional sequence for teaching students about the topics can be very similar.[3] The topic measurement provides many opportunities to teach *via* problem solving.

Possible Instructional Sequences

After defining the characteristic that is to be measured, the major parts of the teaching sequences for each measurement topic can include the use of body units, arbitrary units and finally standard units.

Definition of Characteristic
The characteristics that are to be measured need to be defined for students. Tasks that involve *sorting* and *classifying* are well suited for this purpose. Activities that involve *comparing* and *ordering* make it possible to introduce the appropriate terminology for the characteristic to be measured.

Students need to learn that the commonly used words *big* and *small* are inadequate as well as inappropriate for making comparisons of characteristics other than area. For example, the word *big* does not suffice for a task that requires the selection of the *big glass* from two choices, one tall and narrow, the other short and very wide.

Body Units
After being able to compare and order, students learn to find the answer to questions that ask *how much* according to: *how long; how big; how much does it hold; how heavy;* and, *how long does it take*. The answers to these questions can be found by using units to measure. If these units are *body units*, they are not only developmentally appropriate for young students, but these units make it possible to provide a historic perspective that could have evolved and may be of interest to young students.

As part of these type of settings students will learn such skills and ideas as placing units correctly, using one unit rather than many units, and measuring to the nearest unit using the descriptor *about*. During this stage young students should be encouraged to take risks and make guesses which are valuable pre-requisites for activities involving estimation.

Arbitrary Units
Activities or problems are presented to get students to recognize a need for arbitrary units that are the same for everyone in the classroom. The major skills and ideas for arbitrary units are similar to those suggested for the sequence with body units: placing many units correctly, using one unit rather than many units, and measuring to the nearest unit using the descriptor *about*. Having students design their own instruments to measure can accommodate important learning outcomes. Rather than make guesses, students learn estimation strategies.

Standard Units

Problems are presented that get students to conclude that in order to be able to communicate, it is necessary to adopt standard units that are the same for everyone. Problem settings will give students opportunities to learn how to measure to the nearest unit, see the need for other units, discover the relationships between different units, and become familiar with units that will facilitate the ability to estimate.

Length, Distance and Height

Definitions

The goal is for students to realize that *length*, *height* and *distance* include two endpoints and everything between these two endpoints. To make students aware of this definition, the initial sorting tasks for length and height ask students to consider two categories:

> *Who is as tall as the student at the front and who is not as tall?*
> *Which pieces of string are as long as the stick and which pieces of string are not as long?*

The sorting tasks are then changed to include three categories:

Shorter   – As tall as or the same height   – Taller
Shorter   – As long as or the same length   – Longer

As two students of the same height and two objects of the same length are compared, comments focus on the fact that *height* and *length* include the two endpoints and what is between these points.

To introduce students to *short, shorter, tall, taller, long* and *longer* the following settings can be created. Two used pencils of different lengths are displayed. The longer of the two is labelled as a *short pencil*.

> *What word can be used to describe the other pencil?*

Two students of different heights face the group. The shorter of the two is labelled as being *tall*.

> *What word can be used to describe the other student?*

Two pieces of string of different lengths are shown. The shorter of the two is labelled as being *long*.

> *What word can be used to describe the other piece of string?*

A setting similar to the previous is used with three pencils, three students and three pieces of string to get students to suggest the usage of:

- *short, shorter, shortest*
- *tall, taller, tallest*
- *long, longer, longest*

Wide, wider and *widest*: The students are told that a truck carrying lumber has a sign that reads: *Wide Load*.

> *What word and phrase can you use to describe and compare the load of a truck towing a trailer with a boat on it that sticks out farther than the lumber?*
> *What word and phrase can you use to describe and compare the load of a truck towing a house that sticks out farther than the boat?*

Narrow, narrower and narrowest: Three strips of paper or ribbons, about 10 cm, 5 cm and 1 cm wide, respectively are shown to the students. After the students are told,

> *Consider the 10 cm strip as being narrow, what word can be used to describe and compare the other two strips?*
>
> *What word can be used to describe and compare the 1 cm strip to the other two strips?*

Thick, thicker, thickest and thin, thinner, thinnest: One of several pieces of wood or books of different sizes and thicknesses is labelled as being thick. *Thicker* pieces are pointed to and the students are asked to describe the pieces. As the thickest piece is pointed to, the students are asked,

> *What word would you use to describe this piece?*

The use of pieces of different sizes and thicknesses can illustrate to students the need for descriptors other than *big*.

The request can be made to select the *big* book from a collection on a shelf that includes a very tall and thin book and a book that is not tall but has many pages.

> *What is the problem?* or, *Why is there a problem?*

A similar procedure can be used after identifying and labeling one of the members of a collection as being a *thin*.

Types of Activities and Problems

The activities that are described can lead students to conclude that a statement like, *I am learning to measure* does not make sense or is incomplete. It needs to be followed by the characteristic that is measured.

- **Sorting Words**
  The students are asked to look at words from one paragraph; a spelling list; or at the name tags for the students in the classroom and to identify from the list two members that are the *same length*. The remaining members of the list are sorted according to *longer than*, the *same length* or *shorter than* the two original selections.

- **Ordering Words**
  The students are asked to select a word they consider to be a *long* word. The challenge is presented of finding a word that is *longer* and a third word that would be the *longest* of the three.

- **Sorting Animals**
  A list of names for wild animals, farm animals or a combination thereof is presented. After selecting the names of two animals that are about the *same height*, the remaining members of the list are sorted according to *shorter than*, the *same height* or *taller than* the initial selections.

- **Ordering Animals**
  If a horse is a *tall* animal, select a name from the list of animals that is *taller* and one name of an animal that would be the *tallest* of the three. *What three names of animals could you use to illustrate short, shorter and shortest to someone? Explain your thinking.*

(cont'd next page ...)

■ **Using the Eyes**
A sketch of a part of a door with a **3 cm** by **25 cm** rectangular slot is shown on the chalkboard or the slot is cut from a cardboard box.
The request is made to,
Examine the books on a shelf and try to identify:
- *one book that you think will easily fit through the slot;*
- *one book that you think is almost as tall as the length of the slot and almost as thick as the slot is wide and will fit through the slot;*
- *one book that you think is a little bit too long or a little bit too thick and will not fit through the slot.*

Several students are given the opportunity to select the books they have identified from the shelf, test their ideas and explain how they made their decisions.

■ **Following Directions**
Students are requested to,
- Prepare a sketch of two houses that are far apart.
- Prepare a sketch of two houses, one narrow and one that is wide, that are close together.
- Sketch four houses that differ in widths and are the same distance apart.
- Prepare a sketch of two tall trees that are far apart and two short trees that are close together. Add to the sketch two trees that are shorter than the tall trees and taller than the short trees.

Compare your sketches with that of a classmate. Talk about how the sketches are the same and how they are different. What possible difficulties do you think younger students who know only the words *big, bigger* and *biggest* might have as they try to talk about the objects in these sketches?

Even after many settings that involve comparisons and putting objects in order according to length and height, there will be students who cannot resist the temptation to make use of the words *big, bigger, biggest* and *small smaller* and *smallest*. Patience will be required and frequent reminders may be in order.

Body Units

The attention and focus switches from making comparisons and ordering to finding the answers to:
- How long is something?
- How tall is something?
- How far apart are two things?

Students can be asked to use their *imagination* and consider questions like:
Why would people of long ago want to find the answers to such questions?
What might be some examples of problems they would want to solve?
How do you think they found the answers to these questions?
How do you think they might have measured length, height and distance?

Placement of Units

Students are asked to pretend that they live a long time ago. How might these people use their thumbs or hands to describe the lengths of several selected objects from the classroom? Students are told that what is being used to measure the length is called a *unit*.

Students could be asked whether they know something that is still measured in 'hands' units. Someone might know about reporting the height of horses.

> *How do you think a hand is used or how do you think a hand should be used to measure the height of a horse? Explain your thinking.*

Choosing and Placing Units

A list of objects whose lengths are to be measured is presented. The students are requested to select either a thumb or a hand as a unit of measurement of length and to defend their choice.

A demonstration should invite students' reactions to correct placements and incorrect placements of units. It is advantageous to complete the first tasks with a partner or in a small group setting since this makes repeated placement of the units much easier.

The problem is posed of how the length of something can be measured if friends are not available to lend a helping 'unit.' The correct repeated placement of a unit and making appropriate marks is demonstrated.

To the Nearest Unit

Students need to know what to do when the last unit is placed. They will learn why the word *about* is used since measurement is not exact or it is very rare that units fit exactly onto the objects whose characteristics are being measured. A rule has to be used that allows for consistency in reporting answers when most of the unit is used, when one-half of the unit is used or when less than one-half is used.

Before students measure the lengths and heights of a few objects they should be asked to make a guess about the answers. The intent is to invite risk taking without making any judgmental comments about the responses. The students should realize that all guesses are accepted as being appropriate guesses.

Designing an Instrument

Each student is given a cardboard strip. The request is made to mark and cut off a piece about twenty thumbs long. The strip is used as a ruler to measure length to the nearest thumb.

The students are requested to measure the lengths of two items from their desks and report the result using the term *about*. Students are invited to respond to questions like,

> *If the left side is used as a starting point and the first mark is labelled 1, what does this one mean?*
> *If the next mark is labelled 2, what does the two mean?*
> *In order for a ruler to become a 'counter of units' or do the counting of the number of thumbs for measuring the length of something, what rules need to be followed?*

(cont'd next page ...)

Correct and incorrect examples or non-examples can lead students to conclude that the ruler counts how many thumbs long something is when the left side or **0** is used as a starting point and when the lengths of the objects are placed against the ruler rather than at an angle.

Types of Activities and Problems

- **Using Four Eyes**
  The first experience with using a *referent* and estimating, without actually using the term, can consist of working with a partner, looking around the room and trying to prepare a list of a few objects that are about twenty thumbs long. The lists are compared and the strategies students used are discussed.

- **Using Two Eyes**
  One-half of the ruler that has been made, or a length of **10** thumbs, is shaded. This could be done on the back of the cardboard ruler that has been constructed. The students are asked to keep looking at the shaded part as they try to find and list several objects from around the room that are about as long as this shaded part. The list is compared with the list prepared by a classmate and responses are checked.

- **Drawing Paths**
  With the rulers out of sight the students are challenged to try and draw on a piece of paper a path that they think is about **5** thumbs long and another path that is about **10** thumbs long.

  After the attempt, students are invited to share what they were thinking as they drew the two paths. The rulers are then used to make the paths the requested lengths. This type of task is repeated at a different time. Students can be invited to react to how and why their responses may differ as the task is repeated.

Arbitrary Units

The teacher and a parent who is helping in the classroom take part in completing a measurement of length task that is assigned to the students using a thumb or a hand as a unit. The results from the adults and from several students are recorded on the chalkboard.

Why do the answers differ?
What could be done to have everyone in the classroom arrive at the same answer?
What is one possible problem when a hand is used as a unit to measure the height of a horse?

The adoption of a small unit such as a paper clip makes it easy to reinforce the measurement skills and ideas that have been learned. These skills and ideas include:

- the repeated use of one unit rather than the placement of many units.
- measuring length to the nearest unit. If a 'paper clip ruler' is constructed, a small mark can be added at the half-way mark between two numerals as a reminder, or to make it easier to go to the nearest unit.
- using the word *about* when results are recorded.

Since the types of activities suggested for the use of body units to measure length, height and distance contribute to *visualization* and *measurement sense* these types of tasks are appropriate for arbitrary units. The activities should conclude with a discussion that reminds students of the reason for using something like a paper clip as a unit and then guiding students to discover the inadequacy of using a unit like this.

How would it be possible to communicate with students in other schools, cities, provinces or countries?

Types of Activities and Problems

■ **Look, List and Compare**
To enhance *visual imagery* and to introduce students to the use of a *referent*, they are asked to look around trying to identify several objects they think are about **5** paper clips and about **10** paper clips long. As part of comparing the lists with a classmate, the students are asked to share how they decided on the members of their lists.

■ **Estimation – Strategies**
A path about seven paper clips long is shown on the chalkboard. The students are asked,

About how many paper clips long is the path?

The setting can be used to explain to students or to remind them of the difference between guessing and estimating. When the first thing that comes to mind is reported, this is called making a guess. *Estimating* or *making an estimate* means that a thinking strategy is employed.

Estimation strategies can include:
• Pretending to measure the length of a path by pretending to place paper clips onto the path.
• Comparing the length of a path to something that the measurement of length is known for.
• Pretending to measure the length of part of a path, like one-half, and then using that to report an estimate. To encourage students to use this strategy, it helps to have them estimate the lengths of paths that clearly show two or even three parts and ensuring that it is easy for students to relate these parts of the paths to the total length of the path by using simple addition or multiplication.

Accommodating Responses
If paths are shown on the chalkboard an equity issue exists since all students are unable to look at the paths in the same way. Students sitting at the back of the room or those looking at a path from an angle will see things quite differently than students who are sitting right in front of the drawings. Students should be permitted to get out of their desks, or the drawings could be prepared on pieces of paper which makes it possible to take them to the students' desks.

Standard Units

The idea of being unable to communicate results of measurements with arbitrary units to other people leads students to conclude that units need to be adopted that are the same for everyone.

Metre or Centimetre?

Introducing students to the centimetre rather than the metre, enables them to review and practice the basic measurement skills they have learned:

- Constructing and interpreting the markings and numerals on a measuring instrument.
- Measuring to the nearest unit.
- Reporting results using the word *about*.

Beyond these skills, the main goals for the standard units include:

- Becoming familiar with the units.
- Connecting the units to events and actions from experiences.
- Developing estimation strategies.

Centimetres

Students can use the width of one of their fingernails as a reminder of one centimetre as well as a possible *referent* for tasks that involve estimation.

Most people in the world use the 're' spelling, or the SI (System Internationale) spelling for centimetre. However, the people in the United States use 'er' endings. Students need to be told that the abbreviation **cm** is singular as well as plural and is used without a period and only with numerals.

- **About 1 cm**

  To become familiar with the centimetre as a unit for measuring lengths, students are asked to look in their desks, around the room and think about things from their homes and prepare a list of things that they think are about or close to one centimetre long or one centimetre thick. The lists are compared.

  How are the lists the same? How do they differ?

  A master copy can be prepared for display purposes: **About 1 cm long**.

- **About 10 cm**

  The students can each prepare a similar list and display for things or objects that students think are about ten centimetres long. Occasional discussions and inspections of such lists contribute to becoming familiar with the unit.

  There are no advantages to introducing students to and having them use *decimetre* as a unit. It is not a common unit and its introduction can and will confuse some young students. However, a **10 cm** shaded part of a ruler or a member from the list can serve as a *referent* for estimation activities.

- **Centimetre Ruler**

  The initial ruler that students use should only show marks and numerals at one centimetre intervals. A small mark half-way between the centimetre marks could be added. The activities that have students measure the lengths of objects from their desks and from around the room to the nearest centimetre will result in students becoming familiar with the unit.

(cont'd next page ...)

- **Estimation Strategies**

  Increased familiarity with a unit contributes to being able to estimate with that unit. Estimation tasks should continue to conclude with discussions about the strategies that were employed to arrive at an estimate:
  - *Did the students use a known length to arrive at an estimate?*
  - *Did the students pretend to measure the length?*
  - *Did the students think of two or more parts to arrive at an estimate?*

  Familiarity with the unit and ability to estimate can be fostered by students knowing their heights in centimetres and asking students to be on the lookout for examples of:
  > *Who estimates or measures lengths, distances or heights in centimetres?*
  > *When? Why?*

  The answers to these questions can be used for a display in the classroom.

- **Perimeter**

  Students are asked to suggest how they would estimate the distance around a book or the *perimeter* of a book, and then to try and think of at least two different ways of measuring and calculating this distance.
  > *How do the suggested estimation strategies differ?*
  > *What is the same about the strategies?*
  > *How do the suggested calculation strategies differ?*
  > *What is the same about the strategies?*

  The task of trying to design as many different rectangular shapes as possible for a given perimeter or length, i.e., **16 cm**, can be presented as a problem to be solved with a partner.
  > *What is the same about all of the shapes? How are they different?*

Metres

The students are told that a length, height or distance of **100 cm** is called one *metre*. Students are asked to think of and prepare a list of as many things as they can see and as they can think of that are about **1m** long.
> *Where have they heard of metre being used to measure length, height or distance?*

The entries of the lists and the answers to the question can be used for a display in the classroom.

- **Connecting**

  Familiarity with *metre* as a unit is fostered by connecting the unit to personal actions and experiences:
  - *How many of your steps does it take for one metre?*
  - *About how many metres high is the door?*
  - *About how many metres wide is the classroom?*
  - *About how many metres tall are your parents?*
  - *About how many metres can you jump?*
  - *About how many metres can you throw a ball?*

- **Estimation Strategies**

  Problems to be solved with a partner:
  - *How would you figure out about how many steps you might take in a 100 m dash?*
  - *How would you estimate the distance in metres around the outside of the classroom or the gymnasium?*
  - *How would you estimate the distance around the school?*

  One very valuable part of the problem solving settings that are suggested lies in having students listen to each other as they describe and compare the strategies they would use.

Assessment Suggestions

As students describe objects and discuss strategies, assessment data can be collected about answers to questions of the following type:

- Do students use the appropriate terminology when they compare and order objects according to length, height and distance?

- Do students know why body units or arbitrary units like paper clips are inappropriate for measuring lengths, heights and distances?

- Do students know why the word *about* is used when recording measurements of lengths, heights and distances?

- Do students know what the numerals on a ruler mean?

- Do students know why there is a mark half-way between two numerals on a ruler?

- Do students know the rules that need to be followed for having a ruler tell how long an object is?

- Can students tell why *centimetres* are used to measure length, height or distance rather than thumbs or paper clips?

- Can students identify an object that is about one *centimetre* long or thick?

- When students are given three paths that require 'going up to' and 'down to' the nearest unit and are asked to measure their lengths in *centimetres*, do students go correctly to the nearest unit? Are they able to explain reasons for rounding up or down?

- Do students have at least two *estimation strategies* at their disposal for estimating the length of a given path?

- Can students explain the length of one *metre* in *centimetres*?

- Can students name at least two objects that are about one *metre* long or tall and explain how they made the decisions to select the objects?

- Can students *connect* and tell who uses *centimetres* and *metres* to measure lengths and explain when these units are used?

- Do students' responses show indicators of being familiar with *centimetres* and *metres*? For example, are students able to select the appropriate choice, explain why they think it is appropriate and why the other choices are not?

  How tall is a pop can?   **1 cm**   **10 cm**   **100 cm**
  How high is a door?   **1 m**   **2 m**   **20 m**

- Do students' responses show any indicators of *confidence* and *willingness to take risks*?

Capacity (Volume)

The idea that things might be different from what they look like is true for capacity since three dimensions are involved. Containers that look very different can hold the same amount and containers that look similar may hold different amounts. The idea that it may not be safe to trust one's eyes and that a way of measuring the amount containers can hold is needed. This can be illustrated with a few especially selected containers.

The consideration of three dimensions and how they relate to one another is beyond the scope for students in the primary grades. However, the questions:

Which container holds more?
Which container holds less?
How much can a container hold?
How much is in a container?

are related to students' experiences. Some of the ideas and skills students acquired while learning to measure length, height and distance can be applied to finding the answers to these questions.

Definition

Students can examine several glasses of the same size that are filled with different amounts of water. One of the containers is almost full, another almost empty, one half full, and two with almost the same amounts. The students are requested to provide answers for:

Which glass is fullest or holds the greatest amount of water?
Which glass holds the least amount of water?
Which glasses hold about the same amount of water?

A spoon or a piece of metal is used to slightly tap each glass in random order. The students are asked to help with putting the glasses in order from holding the least amount to holding the greatest amount of water. The glasses are tapped again but this time in order. Students are invited to say something about the two sequences of sounds,

What do you think was different about the two sequences of sounds?

To assist with auditory memory, the glasses are tapped again in random order and then in order and the question is repeated.

What do you think is different about the two sequences of sounds?

Students are faced with a collection of containers that have different shapes and all but two or three differ in size. A bucket of water or sand, or a bag of rice is made available. A glass is presented and the students are asked to identify containers that they think can hold more than the glass.

How is it possible to check that these containers can hold more than the glass?

Guidance may be required to have students conclude that if a glass fits into a container, the container can hold more than the glass.

A similar procedure is used as students try to identify the containers they think can hold less than the glass.

(cont'd next page ...)

Depending on the collection of containers, it may be possible to have students do a three way sort: hold less; hold more; hold about the same amount as the glass.

A collection of containers can be used for ordering them from 'can hold the least amount' to 'can hold the greatest amount' of liquid. The sorting and ordering activities with a collection of containers can lead students to the conclusion that two different looking containers can hold the same amount and containers that look similar may not be able to hold the same amount or may not be the same size. These conclusions lead students to see the need for a way of measuring how much a container can hold.

About Questioning

During one lesson, the students in grade two were looking at two containers of slightly different sizes. The larger of the two containers was filled with water. The students were asked,

How might it be possible to identify the larger of the two glass containers?

After one student suggested that the water from the full container be poured into the other container, one student was invited to do that. After the invitation, there was an uninvited and somewhat impulsive, *'don't spill any'* remark from one of the students. The water was transferred, without any spilling, and the teacher happened to ask,

'Which of the two glasses holds more water?'

The student who was selected to give an answer identified the smaller of the two glasses, and he was correct. The smaller container was full and the other container had very little water in it. A rephrasing of the question resulted in having the student identify the other glass.

The example illustrates that care is required as questions are posed. In this case, there was a great difference between the answers to the questions,

Which glass holds more water? and, *Which glass can hold more water?*

Body Unit, Arbitrary Unit and Standard Unit

Body Unit

How might people of long ago talked about and described how much a container can hold? It could be that they used the unit 'handful' to measure the size of containers.

Students are shown an example of 'one handful' of sand or rice by the teacher and they are asked to record a guess for 'about how many handfuls' they think might fit into two or three of the containers.

After counting and recording the number of handfuls for each of the containers, one container is emptied. A student is asked to fill the container and the number of handfuls it takes to complete the task is counted.

Why is the answer different from before? Do you think the answer is incorrect?

The students are invited to respond to,

What are the disadvantages of using handful as a unit?

(cont'd next page ...)

The conclusions that the 'handful' can be more or be less each time it is used and that it can differ from one person to the next gets students to see the need for the use of a unit that circumvents these problems.

Arbitrary Unit

Students can be invited to respond to,

> What do you think you could use as a unit to find out how much can fit into a container that would give the same answer for everyone in the classroom and would be the same each time it is used?

A small glass jar or tin can serve that purpose.

Many experiences with filling containers and counting how many of a unit it takes to fill them are required before students will be able to come up with estimates. Initially young students should be requested to record guesses. Once the answer for a container is known, this answer can be used as a *referent*. Students' estimates can then focus on two things:

> Do you think another container holds more or less of the unit than the one you know the answer to?
> How much more (or less) do you think it will hold?
> Explain your thinking.

As was the case for measurement of length, the problem of not being able to communicate and share meaningful results with other people is used to get students to see the need for a standard unit.

Standard Units

Students are faced with a collection of different looking containers that hold *one litre*. Any labels that indicate how much the containers can hold are covered. As the students look at these containers, they are asked to respond to,

> What is different about these containers?
> Does anyone know anything that is the same for all of the containers?

Students are told that the amount of liquid each container can hold is the same and it is one litre. Questions of the following type are posed:

> How could you show that these containers can hold the same amount of liquid?
> Which containers in your home have labels that show litres?
> What things are measured in litres?
> Who uses litres? When? Where?
> Why do you think L rather than l is used as an abbreviation for litre and litres?

Assessment Suggestions

Aside from indicators of *confidence* and *risk taking*, assessment data can be collected about three main ideas:

- Do students understand the need for a standard unit?

- Do students know where, when and by whom the standard unit is used?

- Do students show some familiarity with the standard unit?

- Why is the unit 'handful' an inappropriate unit for measuring how much containers can hold?

- What are some things that are measured in litres?

- Select the answer and explain why you think the other choices are inappropriate:

| Ice cream bucket: | **4** litres | **40** litres | **400** litres |
| Milk in the fridge: | **1** litre | **10** litres | **100** litres |

Mass - Weight

Mass refers to the amount of matter in an object and weight is a measure of the pull of gravity on an object. An astronaut's mass is the same on earth and on the moon, but the weight would be different.

Definition

The students are asked to think of an object from the classroom or school that they would consider as being *heavy*. After these objects are named, they are to name an object that they know is *heavier* than the first object. After a student names the pairs of objects for *heavy* and *heavier*, lists of several triplets of names are recorded on the chalkboard, i.e., chair, book, desk. The students are requested to identify the member of the list that they think is the *heaviest* one. After each response the students are asked,

> *Are other answers possible? Explain your thinking.*

A similar procedure can be used to introduce the terms *light*, *lighter* and *lightest*.

After students are asked to think of two things that they think have about the same weight, they are invited to react to,

> *Why is it not always possible or easy to tell which of two objects is lighter or which of two objects is heavier just by looking at them?*

(cont'd next page ...)

Simple sketches of see-saws are shown on the chalkboard. As the sketches appear, the students are asked to tell what they know about the stick people, boxes or the content of the objects on either side of the see-saw:

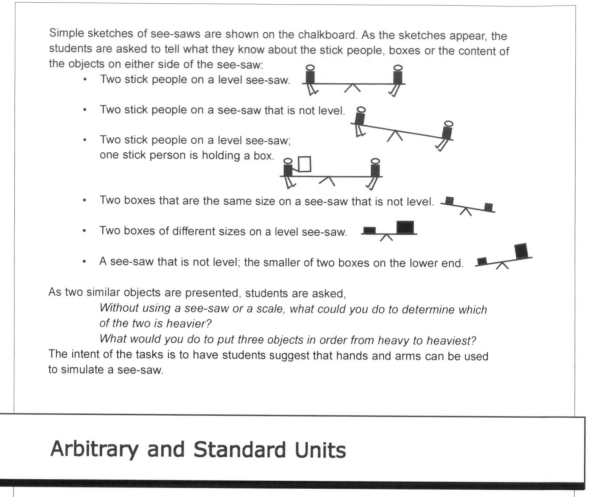

- Two stick people on a level see-saw.

- Two stick people on a see-saw that is not level.

- Two stick people on a level see-saw; one stick person is holding a box.

- Two boxes that are the same size on a see-saw that is not level.

- Two boxes of different sizes on a level see-saw.

- A see-saw that is not level; the smaller of two boxes on the lower end.

As two similar objects are presented, students are asked,

Without using a see-saw or a scale, what could you do to determine which of the two is heavier?

What would you do to put three objects in order from heavy to heaviest?

The intent of the tasks is to have students suggest that hands and arms can be used to simulate a see-saw.

Arbitrary and Standard Units

As a teaching unit about measurement of mass is developed two important questions should be considered:

- Are any important specific cognitive learning outcomes related to measurement skills accommodated by having students use arbitrary units such as blocks or nails to describe the mass of objects?
- Should a teaching sequence begin with introducing students to grams or to kilograms?

Many references for students as well as for teachers do suggest activities that have students use arbitrary units to describe the mass of objects. The references also provide suggestions for the construction of simple balances which, for example, include the use of coat hangers. One thing these references fail to do is provide teachers with specific cognitive outcomes for these types of activities. What do students learn from these types of tasks? Are the tasks important or required for learning about measurement of mass?

The assumption is made that these tasks do not contribute to aspects of cognitive development and therefore it is advantageous to move on to the second question, which deals with the use of a standard unit. The following question can serve as a possible guiding principle,

Which unit, the gram or the kilogram, is more easily related to the experiences of young students?

(cont'd next page ...)

A very delicate balance is required to record results to the nearest gram. It is not very easy for young children and for most adults to tell the difference between one gram and two grams. Grams are not easily related to young students' experiences. These reasons are suggestive of advantages for an introductory sequence that begins with introducing the kilogram.

Kilogram

One introductory activity can involve providing students with a piece of Plasticine or clay with a mass of one kilogram, without the students' knowledge of this fact. The request is made to use all of the Plasticine and to sculpt a shape from the piece that looks different from the shapes the students they are sitting next to are sculpting. The sculptured shapes are displayed and the students are invited to respond to,

> What is different about the sculptures?
> Does anyone notice anything that might be the same for all of the sculptures?

Guidance will be required to have students conclude that all of the pieces weigh the same; that is if nobody removed any parts from the piece of Plasticine. Several students could be asked to hold two pieces in each of their hands to assist with reaching the conclusion that the pieces have the same weight. A balance is used in support of those who agree and to convince those who are doubtful that this is in fact the case.

Students are told that each of the pieces and shapes weighs *one kilogram.*

> By looking at the sculptures, what can you say about things that weigh one kilogram?
> Try to think of something you know that weighs about one kilogram.
> Where have you heard of or where have you seen kilogram used?

What weighs about one kilogram? With a partner make a list of several objects in the classroom that you think weigh about one kilogram. The lists are compared and combined for a display on the bulletin board. The students are asked,

> About how many kilograms do you think it would it take for a see-saw to be level if you are sitting on one side of it?

Five different stacks of books are displayed on a table. The stacks are labelled **A** to **E**. The request is made to,
Try and identify and record the letter for:
  * one stack that you think 'weighs less than one kilogram.'
  * one stack that you think 'weighs about one kilogram.'
  * one stack that you think 'weighs more than one kilogram.'

The students are then invited to think about how they explain their thinking for,

> What could you do to estimate and check to see whether your recordings are correct?
> What would you do to put the stacks in order from weighs the least to weighs the most?

As several students are given the opportunities to share and explain their strategies, the observers are invited to make comments or to ask questions.

Gram

As part of a task, students are led to conclude that the kilogram is not appropriate to measure the weight of many objects that are found in the desks and in the classroom. The *gram* is introduced as a unit that is about the same as the mass of some candies or a stick of gum.

The students are requested to try and list several things that they think weigh about one gram and to suggest answers for,

Who uses grams? When? Where?

The results are compared and a display can be prepared.

After the students are told that **1000 grams** are equal to *one kilogram*, they are asked to try to find their weight in grams. Strategies for finding the answers are discussed and compared.

Assessment Suggestions

Data can be collected about three main ideas:

- Do students understand that there exist many instances when their eyes cannot tell them which of two objects is lighter or which is heavier?

- Do students know where, when and by whom the standard units are used?

- Do students show some familiarity with the standard units?

As students explain their thinking, evidence about *confidence* and *willingness to take risks* may be noted.

- What can students tell about two boxes that differ in size on a level see-saw?

- What can students tell about two boxes with the smaller of the two on the lower end of a see-saw?

- Can students name something that weighs about one kilogram?

- Can students tell what is weighed in kilograms and by whom?

- Can students name something that weighs about one gram?

- Can students tell what is weighed in grams and by whom?

- Can students select the correct answer and explain their thinking:

A pencil:	More than **1** kg	Less than **1** kg
A postage stamp:	More than **1** g	Less than **1** g
A book	About **1** g	About **1** kg
A raccoon:	About **8** g	About **8** kg
A blue jay:	About **75** g	About **75** kg

Time

Time is used to specify when events have or will occur and to describe how long events will last or have lasted. Judging the passage of time can be and is very subjective as indicated by comments used as part of conversations like, *'time seemed to drag'* and, *'time went by quickly.'* There are times when young children will share the observation, *'your minute is too long'* and there is a time period when they actually believe that time passes more quickly and the hands on a clock move more quickly when tasks are done quickly. [4]

Reading clocks with understanding is a complex process. It requires an understanding of the marking on the faces and the functions of the hands. Readiness activities are about sequencing of events and the duration of time. A teaching sequence similar to that suggested for the other topics of measurement can be developed to reach the goal of fostering *conceptual understanding*.

Definition

The students are asked to describe two actions from their experiences that they think *take little time* and two actions that they think *take a long time* to complete. Two actions or events from the students' experiences are described and the students are asked to tell which of the two they think *takes longer* or *more time* to complete.

For example:
        Drawing a picture – hanging a picture.
            Feeding the cat – combing the cat.
      Building a bird house – painting a bird house.
             Brushing teeth – combing hair.

After responses are elicited, the students are asked to consider,
> *Why is it possible for some of these actions to have different answers for different people?*

Students are challenged to try and think of two actions that take them about the *same amount of time* to complete. It may be necessary to provide a list of possible examples, i.e., opening and closing a door; getting into and out of a car; putting on a shirt and putting on a jacket. As the ideas are presented, the listeners are asked to think about themselves.
> *Do you think one task takes more time to complete and the other less time or do you think both tasks take about the same time to complete?*
> *Be ready to explain the reason for your decision.*

Three events or actions are described and the students are requested to put these in order from *takes the least amount of time* to *takes the greatest amount of time* to complete.

For example:

Brushing teeth	Combing hair	Washing hands
Lunch at school	Recess	Mathematics class
Favourite TV show	Art class	Taking a bath
Eating lunch	Eating breakfast	Eating supper

Students are invited to respond to,
> *Why is it possible for the order to be different on different days?*
> *Why is it possible that the order is different for different students?*

'Body' and Arbitrary Units

It is easy to decide which of two tasks takes longer to complete. Sometimes people are interested in how much time it takes to complete a task and in how much more or less time one task takes than another. How might it be possible to determine how long an action takes without the use of a watch or clock?

Students are told that deciding how long something takes could be determined by clapping hands or tapping a pencil and counting the number of claps or taps. To illustrate this method of timing, several actions that can be performed by students in the classroom near their desks are suggested.

For each suggestion that is made, the students are requested to record a guess for the approximate number of *claps* they think it might take to complete the task. After the students have recorded their guesses, the signal to start is given and the clapping and counting begins. The students record the number name they hear when they have finished the task.

Possible tasks:	Guess	Actual Count
Printing my first and last name		
Printing the letters of the alphabet		
Standing up twenty times		
Reading one page		
Printing the number names for **1** to **10**		
Printing my address two times		
Printing my phone number twice		
Sharpening a pencil		
Getting dressed for recess		

To assist the students with the conclusion that clapping and counting is not appropriate to tell how long something takes, the rhythm of clapping is changed for one or two of the tasks. The students are invited to state possible disadvantages of using this method of determining how long a task takes, other than *'the clapping is too loud'*, as one student in grade three concluded.

The students are introduced to a metronome which is set for clicks one second apart. However, the students are not told about this one second setting. The important part now is that the counting will be done at the same pace, and at the same noise level!

Questions need to be posed that lead students to conclude that they were able to find out how long an action took, but people outside the classroom would not be able to get the same answers since they do not know how quickly the clapping was done or how quickly the clicks of the metronome happened.

Standard Units

The sequence of suggested activities moves from the smallest unit, one second, to the largest unit, one day. In some references that sequence is reversed. No matter what the sequence, *conceptual understanding* should be the major goal and the adoption should be based on which of the two sequences might best be suited to reach this goal.

One Second

The students are asked to listen to the time between two clicks of the metronome and they are told that this time is equal to *one second*. The students, with a partner, are requested to think of things that they think would take about *one second* to do or things that could be done in one second. The lists are compared.

Students are invited to respond to,

When have you heard people use one second as part of their comments?
Who might be interested in seconds and why?
What are some things that are measured in seconds?
What word or number name can be said in one second?

While visiting schools in Great Britain, a Canadian science teacher heard a member of a group of students use the word *Mississippi* to time swings of a pendulum. He told the group of students that rather than *Mississippi* in the part of Canada where he was from students use *Saskatchewan*. A little later, one student of the group was overheard saying, '*Saskatche-one, Saskatche-two*', etc.

No doubt almost every student has heard from someone the comment *I will be with you in a second.* The students are asked,

How would you respond to a person who makes such a statement?
Why would you respond in this way?

One Minute

The students are told that a count to *sixty seconds* is equal to *one minute*. A discussion could begin with,

What do you know about a minute?
What do you think you could do in one minute?
When have you heard someone use the word minute?

Answers to the last question will likely include such comments as,

I will be with you in a minute.
Just wait one minute.
It won't take long, just another minute.
You have one more minute and then we have to go.
I will give you just one more minute.

What do you think the people who utter statements like these really mean? Why are these comments difficult to interpret? Persons who utter these statements may have something other than a minute in mind and they may not check when that one minute is up.

The duration of one minute is not meaningful to young students. They do not have some sort of personal *referent* for this unit of time.

(cont'd next page ...)

One of the major goals is to connect this basic unit of measurement of time, *one minute*, to students' experiences. This connection will enable students to attach some sort of meaning to statements that make reference to this unit. One of the best ways to reach this important goal is to have students construct their own *One Minute Booklet*. The content of the booklet lists actions that students can complete in this period of time. The entries will enable students to state what one minute means to them in terms of actions familiar to them.

The activities and actions listed in the booklet can be part of different subject areas, i.e. handwriting/printing, language, reading. Activities from the gymnasium, during recess or out on the school ground can be included. The headings in such a booklet could be:

How many can you complete in one minute?

Action	Guess	Actual Count

Students who know how many times they can: skip rope; print their name; print the letters of the alphabet; read a page in a book; sing a verse of a favourite song; bounce a ball; etc. in *one minute* will have a *personal referent* at their disposal that will enable them to attach some sort of meaning to statements that require them to think of this basic unit of measurement of time.

Students can be asked, while working with a partner, to think of and record scenarios when people they know use *one minute* as part of their comments, requests or commands. These scenarios could be shared with classmates as part of brief acts or pantomimes.

> *Did the people who uttered these comments have one minute in mind?*
> *Why or why not?*

The idea about of how long a minute is takes time to develop. In order to accommodate the notion, students should be given the opportunity to add personal experiences to their *One Minute Booklet* throughout the year.

One Hour

A number line from zero to sixty with sixty marks is displayed on the chalkboard. Every fifth number name from **0** to **60** could be shown on this number line. The students are told that the distance between two marks on this number line indicates a one minute time interval. Reference can be made to the *One Minute Booklet* and the number of times certain tasks were completed during this interval.

A vertical arrow is moved along the number line to indicate the time that passes for events from the students' experiences. These events are recorded on the one hour number line. For example, the arrow starts at **0** and stops at **15** for the response to:

> *Where should the arrow stop at the end of recess or after fifteen minutes have gone by?*

An R or the word Recess is recorded below the **15**. A similar procedure is used for other events. For example,

> *Where should the arrow stop to show how much time is set aside for lunch?*
> *Where would the arrow stop to show how long your favourite TV show is?*

With a little imagination, the students could think of the number line as a very primitive watch. What if a cardboard number line was tied to one wrist and with the other hand an arrow was moved to show how much time has passed by? What are some possible disadvantages of such a watch?

No wonder someone thought of putting the number line in the shape of a circle and designed a long arrow or a long hand to show the passing minutes. It is easy to tell just by looking at the big hand when, for example, **5**, **12**, **37**, or **50** minutes have gone by.

*Where would the big hand point to after **67** minutes have gone by?*
Explain your answer.

What is needed is another number line that shows how many times the long hand has gone around the circle or how many hours have gone by. Rather than having to use two watches, or one for each wrist, another number line and a short arrow or short hand is added to the watch. This short hand tells how many times the big hand has moved around the circle or how many hours have gone by since both hands started at zero.

Telling Time

What are people asking when they pose the question,

What time is it?

People who ask that question want to know the answers to,

What number names do the two hands point to on the two number lines?
How many times has the short hand moved around and what number name is the short hand pointing to?
How many minutes have gone by or what number name does the long hand point to?

For example, when the long hand points at the **0** or the **12**, and the short hand to the **3**, at one time people said, *It is three of the clock* or, *It is three o'clock* which means that the long hand has gone around three times or three hours have passed since both hands pointed to the **12**. A digital watch would show **3:00** which means that the hour or short hand points to the **3** and the minute hand or long hand to the **0**. Thirty minutes later, the short hand has moved half way between the three and the four. The long hand would point to the **30**. The answers to the question, *What time is it?* could be:

It is thirty minutes after three o'clock or, *Three thirty.*
It is thirty minutes to four o'clock.
It is one-half hour past three o'clock or, *It is half past three.*

A digital watch would show **3:30** which means that the short or hour hand points to **3** and the long or minute hand points to **30**.

One Day

When the short or hour hand moves around the number line once, twelve hours will have gone by. After going around the second time, twenty-four hours have passed. A twenty four hour period is called one day.

Assessment Suggestions

As students describe actions and discuss strategies assessment data can be collected about answers to questions of the following type:

- **Need for Standard Units**: What is a disadvantage of clapping and counting to measure how long it takes to complete a task?

- **One Second**: Name something that you think takes you about one second to do. About how many seconds do you think it would take to print your name? Choices could be presented:

  About **1** second   About **10** seconds   About **50** seconds

- **One Minute**: How many seconds does it take for one minute? How could you try to figure out when one minute has gone by without looking at a watch? What else could you do?

- **One Hour**: How many minutes does it take for one hour? Name something that takes about one hour.

- **Telling Time**: If you are told, *'It is five o'clock'*, what number names would the hands on a clock point to and what does that mean?

Reporting

During presentations to parents of preschool children and students in the primary grades, there will be some who make the comment that they are teaching their young children about measurement while they are cooking and baking with them. Most of these parents are not aware that the ideas and skills related to aspects of measurement are complex and involve highly complicated procedures.

For many parents it will be informative if they are told what the major general goals for learning about measurement are. An awareness of these goals cannot only lead to meaningful dialogues between them and their children, but the questions posed and comments made during activities that involve aspects of measurement can become more meaningful.

After informing parents that the topics of measurement deal with length, distance, capacity (volume), mass (weight) and time,

they can be made aware of the key ideas that are or were part of the teaching activities. These include:

- Use of appropriate language for each characteristic that is measured as comparisons are made.
- Use of appropriate language for each characteristic that is measured as three or more objects are ordered.
- Learning to measure some characteristics to the nearest unit and learning the use of the word *about*.
- Being able to explain the difference between estimating and guessing.
- Being able to explain the meanings of the markings and numerals on instruments that are used to measure length and time – the ruler and the face of a clock.
- Knowing who is interested in measuring length, distance, capacity, weight and time and knowing why standard units are used: centimetres, metres, litres, kilograms, and grams.

An awareness of the key ideas can result in settings and conversations in the home that reinforce what has been learned in the classroom.

For Reflection

How would you define *measurement sense* and what indicator of *measurement sense* could you use to illustrate your definition?

What main points would you include in a presentation about: *Measurement Sense: Guessing and Estimating – Building Confidence in Students*?

How would you respond to a parent who asks, *Why do you have your students use body units and arbitrary units to measure characteristics of objects?*

What suggestions would you make to parents who ask, *What could we do at home to foster the development of spatial sense and measurement sense?'*
Is there anything we should not do?

Consider a reaction to the following scenarios:
   A father of a child in Kindergarten states, *'My son knows his numbers. He can count to thirty.'*
   A mother of a child in Kindergarten states, *'My daughter knows how to measure because we cook and bake together.'*

Many of the main skills and ideas that are part of *data analysis* are well suited for teaching *via* problem solving. The activities and problems that are part of data analysis can contribute to fostering students' development of:
- language skills.
- aspects of reading comprehension.
- evaluative skills.
- probabilistic thinking.
- arithmetical skills.
- visualization.

Since the activities related to *data analysis* can make contributions to other important areas of learning, and since *data analysis* is part of other subject areas, it is appropriate to suggest that it is advantageous for students to encounter the topic during the early part of their yearly mathematics learning.

Teaching Sequence – Key Ideas

A possible teaching sequence is described in detail for one sample problem. The intent is to focus on the major ideas, procedures, skills and to illustrate the wide range of specific learning outcomes that can be accommodated by having students collect, record, organize, label, display and discuss data about a question that they do not know the answer for.[1][2]

For example,
While discussing the four seasons, the following problem is posed,
> *Into which of the four seasons do the greatest number of birthdays fall and into which season do the least number of birthdays fall for the students in this classroom?*

Four pieces of paper are labelled:
- Spring – March, April, May
- Summer – June, July, August
- Fall – September, October, November
- Winter – December, January, February

The pieces are placed on the floor and the students are asked to stand around the season into which their birthdays fall.
> *Which season has the greatest number of birthdays?*
> *Which season has the least number of birthdays?*

What could be done to make the answers to these questions very obvious just by looking at the students in each group?

(cont'd next page ...)

The organization of data that have been collected includes decision making as well as problem solving. The learning about all of the possible conclusions that can be drawn from organized data provides many opportunities for mathematical thinking.

Aspects of Organized Data Collections

Equal Spacing
The students are guided to conclude that standing behind one another and having every group do it in the same direction would make it easy to make statements about the answers to the questions about greatest and least number of birthdays. The importance of equal spacing between the students can become part of the discussion if a group with fewer students is aligned with greater distances between the students.

Same-sized Units
Students are led to conclude that since every student is in the line-up and each one is part of showing the answers to the questions, it is not easy for those who are lined up to make comparisons about greatest number, least number, about the same number, or about differences in number. To circumvent this problem, the pieces of paper identifying the four seasons are placed on a table. As students come up to the table they are given a block and they are requested to drop the block onto the appropriate season.

A few of the students are given blocks that differ in size from the remainder of the blocks. After students have dropped their blocks, they are asked to suggest what should be done to make it easy to see the answers to the questions.

> *Do they notice anything that needs to be changed?*
> *If so, why is that the case?*

The students are asked to look at the stacks of blocks and make statements about all of the things they can think of that the display tells them about birthdays in the four seasons. Suggestions may need to be made to have students use:

> *more; fewer; the same number of or about the same number of;*
> *closest or least difference; farthest apart or greatest difference.*

> *How could the display be changed to give information about birthdays*
> *for the boys and girls and the four seasons?*

After the advantages of the block display over the people display are discussed, the students are invited to think of possible disadvantages of the former. One disadvantage has to do with sharing the display with others in other classrooms, schools or cities.

To solve the sharing problem and to make it easier to display and examine results, a piece of paper is mounted on the chalkboard or the bulletin board. The four seasons are printed along the bottom of the sheet. The students are provided with stickers. As they walk past the piece of paper they place their sticker over the appropriate season.

Rather than using stickers, students could use square shaped pieces of paper or shade in squares on graph paper. If a felt pen is used to draw around the perimeter of the rectangular region of the squares that are placed on the piece of paper, a bar picture or bar representation or *bar graph* is created.

(cont'd next page ...)

Titles

If a student from another classroom, another teacher, or the principal walked into our classroom, why would they be unable to tell what the display they are looking at is all about?

> Every display needs a title. Books have titles. Movies have titles.
> *What is a title and what does it tell us?*
> *What is a good title or what makes a title a good title?*
> *What are possible examples of titles that are not good? Why is that the case?*

The discussion should lead students to conclude that good titles tell what something is about. Other characteristics might include 'eye-catching' and not too long. The titles of stories in readers could be examined for some of these characteristics.

Students could be asked to react to suggestions for titles for the displays they have prepared. For example, reactions to the following suggestions for titles for the four seasons and birthdays could be invited:

- **Our Birthdays**
- **Most Students Are Happiest in Spring Because That is When Their Birthdays Are**
- **When Do We Have Our Birthdays?**
- **The Birthdays for Students in Our Grade**
- **Very Few Students Like Winter**

Labelling

One part of the display about the birthdays is almost self explanatory. It is labelled **Seasons**. To get around the bother of having to count the number of labels for every question that makes reference to numbers, number names are listed at appropriate intervals and a label is included as a reminder – in this case, **Number of Birthdays**.

Interpretation Skills

The ability to interpret displays and graphs is a major learning outcome.

Comparisons and Calculations

The simplest types of interpretations deal with aspects of arithmetical skills. Answers to questions that include such terms or phrases as:

> *most; least; the same number of; about the same number; how many more; how many fewer; finding sums for parts of the display; finding differences between parts of the display.*

Inspections

Some questions about a topic that can be answered by looking at a display or graph. For example,

> *Into which season do the most (fewest) birthdays fall?*

Could be True

Some questions may deal with a topic that is part of a display or graph but they cannot be answered by looking at the display or graph. For example,

> *Are Susan's and John's* (names of students in the classroom) *birthdays in the summer?*

(cont'd next page ...)

True, False or Could be True

Students are requested to explain their thinking of their choices as they classify statements about the topic of a display or graph as:

- **True**,
- **False** or,
- **Could be True, but the display or graph does not tell us**.

For example,

- Every student's birthday from the class is included in the display.
- The teacher's birthday is included in the display.
- All the girls in the classroom have a birthday in the fall.
- The display shows how old the students are.
- More boys than girls have a birthday in the winter.
- Students like having their birthdays in the spring.

A Graph is Worth ...Words

Students work with a partner and they are asked to think of all of the different statements they can think of by looking at what a display or a graph could tell someone. All of the different phrases, sentences and conclusions that are then shared should result in a degree of amazement and a conclusion that a display or graph is almost like a picture because it is 'worth so many words'; which is a good reason for drawing graphs and preparing displays.

These types of interpretation questions, conclusions and discussions can be part of any display or graph the students are shown or are constructing.

Related Problems

Every display and graph that is constructed can lead to further problems to be investigated. While displays are interpreted and discussed, students can be asked to think of other questions they might want to find the answers for.

Data Collection – Tally Marks

To find out how the graph for the students in our classroom compares to students from another classroom, data need to be collected. Since these students from the other classrooms are not available to stand beside pieces of paper on the floor; drop blocks onto pieces of paper or put stickers on graph paper, the students need to be asked questions.

A stick person could be drawn to represent each student's response or a simple stroke with a pencil called a *tally mark* can be recorded. These tally marks tell how many birthdays there are in each season and then they can be used to construct a display or a graph.

(cont'd next page ...)

Organizing Collected Data

After data are collected from another group, students could work with a partner and use the tally marks to,

- Construct a display on graph paper.
- Label the parts
- Make up a title.
- Write sentences about your conclusions.
- Make up two questions that can be answered by looking at the graph.
- Make up a question about the topic of the graph that cannot be answered by looking at the graph.
- Make one statement about the topic of the graph that the graph shows is true.
- Make up one statement about the topic of the graph that the graph shows is false.

Students share their graphs, questions and statements.

For discussion:

Do you think the graph will be exactly the same next year? Explain your thinking.
Do you think separate displays for boys and girls in this classroom would be similar or different? Explain your thinking.

Types of Activities and Problems

■ Questions about Ourselves

- What is our favourite recess (Physical Education) activity?
- What is our favourite fruit (breakfast)?
- What is our favourite TV show?

■ Investigations

- Which vowels are used most and least frequently?
- How many syllables do most of the words on one page have?
- With what letter do most of our names start?
- How many letters are there in each of our first (last) names?

■ Measurement

- How many times can we print our first name in twenty seconds?
- About how far apart can we spread our fingers?
- How close did we come to drawing a path 5 cm long without the use of a ruler?

■ Use of Imagination

- Graphs with missing information are presented, i.e., missing title, missing labels. The students are asked to suggest what they think the missing information could be and to explain their thinking.
- A graph with all of the information missing is presented. The students are asked to make up a title and labels for the graph that they think make sense to them. The students share their thinking.

Data collection, organization of data and the interpretation of the data can also become part of game playing settings. An example is included in the next section.

Assessment Suggestions

Students are requested to provide responses to questions and statements about a display or a graph.

Favourite Flavours in the Grade 2 Classroom	
Vanilla	✷ ✷ ✷ ✷ ✷ ✷
Strawberry	✷ ✷
Chocolate	✷ ✷ ✷ ✷ ✷ ✷ ✷
Maple	✷ ✷ ✷ ✷

- Why does the display need a title?

- Do you think this title is a good title? Explain your thinking.

- How many students like chocolate?

- How many students like strawberry and maple?

- Tell whether you think the statements are <u>True</u>, <u>False</u>, or <u>Could be True</u>, <u>but the display does not show it</u>.
  Explain your thinking.
    a) There are twenty students in the Grade 2 classroom.
    b) Strawberry is liked the least.
    c) The display is about ice cream.
    d) Two girls in the classroom like vanilla.
    e) The teacher likes chocolate.
    f) Nobody in the classroom likes raspberry.
    g) Most students like vanilla and strawberry.

- Make up one statement about the topic of the display that is True.

- Make up one statement about the topic of the display that is False.

- Make up one statement about the topic of the display that Could be True, but the display does not show it.

Accommodating Responses

Students' responses may differ because they might be considering the inclusion of the teacher and a teaching assistant or perhaps another adult who might have been in the classroom on that day. Students need to be given the opportunity to explain the thinking behind their responses.

Game Settings

The assumption is made that appropriate game settings can be an important part of mathematics teaching, learning and assessment.

Goals for Students

Game settings that contribute to reaching the goals that are listed in the mathematics curriculum can be created. Exchanges can take place during games that can foster *self-confidence* and encourage *willingness to take risks*. The discussions that can take place while games are played are valuable since they can foster communication skills as well as thinking skills.

Activities related to games can give students opportunities to *communicate* in written form. Many aspects of *mathematical reasoning* can be accommodated in game settings. These can include:

- Classifying information into true, false, and could be true or could be false categories.
- Looking for possible reasons for the outcomes that occurred.
- Discussing moves and possible reasons for moves that might be detrimental.
- Trying to make predictions.

Game settings provide opportunities to *apply* skills and ideas that have been learned and *connect* these to new settings. Game settings can be created and orchestrated in a way that will contribute to the development of a *positive attitude* toward mathematics and to *perseverance*. Questions can be raised and tasks can be assigned that provide opportunities to *exhibit curiosity* and *use of imagination*.

Mathematical Processes

Many of the *critical components* identified in the mathematics curriculum can be accommodated in game settings. *Communication skills* can be fostered. Mathematical skills and ideas are *connected* to new skills and ideas. Opportunities for *mental mathematics* and *estimation* can be provided. It is possible to pose questions or to assign tasks which create a learning *through* problem solving setting. Aspects of *mathematical thinking* which can include: *sense making*; *getting* oneself *unstuck*; *identifying errors* in thinking, in moves or in the use of materials; and opportunities to *try different strategies* can all be accommodated. Some game settings can contribute to fostering the development of the ability to *visualize*.

The many possible positive outcomes that have been identified are supportive of the inclusion of games, or certain types of games, as part of ongoing mathematics teaching and learning.

(cont'd next page ...)

What is a Game? What are Criteria for Appropriate Games?

Since there is something intrinsically motivating about games or a statement like, *Let's play a game,* the latter is used many times for settings that are not games or do not even resemble a game. In fact, quite a few authors of educational materials use the term game rather loosely for all kinds of activities. To be consistent with the students' experiences games should have rules and there should be a winner or winners.[3]

Several criteria for effective or appropriate games are listed. It is unlikely that there exist games that meet all of these criteria, but the assumption is made that the more of the criteria are met, the better the game.

Criteria for effective games can include:

- The rules should be simple. Time required to give lengthy explanations can lead boredom and confusion for young students. Simple rules make it easier to alert students to use their imagination and to try and think about possible rules that might be added in the future.

- There should be quick action. Quick action helps to maintain interest. Slow action and waits between moves or opportunities to participate can result in boredom and loss of interest.

- Few and simple pieces should be required. Money does not need to be invested and little needs to be kept track of. Too many pieces can easily distract from the intended goals of a game. Any loss of pieces or unintentional rearrangement of pieces during a game can result in breaks in the action and a loss of interest in a game.

- The emphasis should not be on winning. Games with a chance outcome give everyone an equal opportunity to win. The emphasis should not be on fighting one another or winning a war. With all the violence, references to war seem inappropriate. An inability to predict the winner can be reinforced by using 'Lucky' as part of the name for a game. Finishers could be called first winner, second winner, etc. and every payer should be rewarded is some way, i.e., awarding **10**, **9**, **8**, **7**, etc. points to participants.

  A chance outcome does not imply that possibilities to think, to solve problems and to ponder strategies do not exist. For some games these strategies will not affect the outcomes of a game. There exist appropriate games that allow those who do solve problems and develop strategies to become better at playing these games.[4]

- The game should be a learning experience. An incorrect response should not be punished in any way. Opportunity should exist to correct and to keep participating.

(cont'd next page ...)

- The game should allow for questions, discussions and responses to certain types of questions during the game. Discussions could focus on invited comments to questions of the following type:
    - *If you did 'this' or made 'this' move, why would you do it?*
    - *What do you hope will happen next? Why?*
    - *I heard someone sigh. Why do you think that happened?*
    - *Some students said, 'Yes!' – Why do you think they did?*
    - *Show on your fingers what number you would like to have come up next? What do you think we can say about students who want that number to come up next?*

- Good games are flexible. They can be used in different settings – large groups, small groups, with a partner or in a solitary setting.

- Physical actions are desirable in some settings, but in the mathematics classroom they can and do distract from the purpose or the outcomes of a game. Many observations have been made where the main attraction was throwing bean bags, tossing rings or casting fishing rods rather than solving problems related to aspects of mathematics.

- Good games can easily be modified to accommodate new topics and new learning outcomes.

This list of criteria is suggestive of the existence of games that are not appropriate or are educationally unsound. They do exist.
These types of games:
- may have too many parts or pieces to keep track of;
- have rules that can be quite complicated and require lengthy explanations as well as examples for strategies to be used during the game;
- may punish inappropriate or incorrect responses by students by taking them out of the game;
- may end with few or just two participants;
- do not let players know during a game whether or not their responses are correct;
- may be considered inappropriate by some groups of people since they make reference to aspects of gambling or involve some sort of gambling.

It is not suggested that the games that are included in this collection of examples are necessarily the best games that are available, but observations collected over the years indicate that these games were well received by students. Indicators of this reception are not just the reactions that were observed while students played these games, but also include the repeated requests to play the games again.

The names for these games are arbitrary. Since students seem to enjoy making up and giving their own names to games they can be given that opportunity.

Guessing, Strategy and Data Analysis
How Many are Hidden?

Setting: Two players.

Materials: Four counters, two for each player.

The goal of the game is to guess the number of counters held in two hands.

Procedure

The players hold the two counters and both hands behind their backs. Two, one or zero counters are transferred into one hand which is then closed, brought to the front and placed onto the table.

One player gets first turn with trying to guess how many counters there are in two hands on the table. Then the other player takes a turn with the restriction that the number guessed by the first player may not be used again.

The players open their hands to see whether or not a correct guess was made and a winner can be declared. If a correct guess was not made, it is called a draw or tie. For the next round of the game the procedure is repeated and the player who guesses last now has the first turn.

The setting is well suited for preparing several frequency tallies or bar graphs to keep a record of the outcomes:
- Which number names were guessed?   **0   1   2   3   4**
- What were the winning number names? **0   1   2   3   4**
- What were the outcomes of the game? **Win   Draw**
- Which guess was the winner? **First Guess   Second Guess**
- Who won? **Names of Players: A   B**

Possible Variations
- Six counters are used. Comparisons can be made.
- Three players play the game with two counters and then three counters. What is the same and what is different?

If the frequency tallies for *Which guess was the winner?* are combined for the whole class, the result will show that there are advantages to making the last guess. Why is that the case?

Number Sense
To and From Ten

Setting: Two players or two teams; the students on one side of the classroom against the students on the other side.

Materials: Two cardboard strips, each marked off into ten rectangular shapes or spaces in a row. The size of the shapes allows for a placement of a button, bottle cap, penny, small block or chip. An envelope with four disks or four pieces of cardboard; each shows one of the numerals **1** to **4**.

The goal of the game is to find out who is lucky to be the first to fill up all of the ten spaces on the game board with chips: **Lucky to Ten**.

Procedure

Players take turns drawing a chip or piece of cardboard from the envelope.
The number name tells how many chips can be placed onto the game board.
The chip or piece of cardboard is returned to the envelope and it is shaken.

To illustrate how the game is played, the students on one side of the room could face those on the other side. The game boards are shown on an overhead or are sketched on the chalkboard. Dots are drawn or magnetic pieces are used to fill the spaces.

The following stage of a game in progress can be used to illustrate the importance of conversations during the game.

Player or team A has seven chips on the game board, and player or team B has four chips on it.

The players are asked to respond to types of requests like:
  * *Who is ahead?*
  * *Use fingers to show how far ahead player A is.*
  * *Use fingers to show how many more player A needs to go to ten or to cover all ten spaces.*
  * *Use fingers to show how far behind player B is.*
  * *Use fingers to show how many more player B needs to go ten or to cover all ten spaces.*

After each turn, a player is asked,
  * *How far behind (or ahead) are you?*
  * *How many more do you need to cover ten spaces?*

Possible response options by players include:
  * showing the number with fingers.
  * showing the number with fingers and stating the number name.
  * stating the number name.

Toward the end of a game a problem needs to be solved.
What should happen in order to cover the last space or spaces?

(cont'd next page ...)

Number Sense
To and From Ten (continued)

Should players wait for the number drawn to match the number of empty spaces on the board, or would a number naming a number greater than the number of these spaces suffice?

The students can decide what they think should be done.

The students can be given the opportunity to make up a name for the game.

After one player was lucky to be first to cover the ten spaces, the first lucky winner, the second player is given the opportunity to cover all of the spaces, and to be the second winner.

The goal of the game can now change to:
Who will be lucky or be the first lucky winner to take off all of the chips?
        Lucky from Ten or **Lucky to Zero**.

The types of questions after every move can now be:
- *Who is closer to having zero chips on the board?*
- *How far ahead (behind) are you?*
- *How many have been taken off?*
- *How many more need to be taken off?*

The students can decide on the procedure to follow for taking the last chips off the game board.

As the game moves along students are requested to ask each other the questions:
- *how far ahead (behind)?* and,
- *how many more to go to ten (zero)?*

Possible Variations
- Different number names can be added to the envelope.
- Some names could be repeated.
- The game board could be enlarged.
- Students could play the game alone. They keep track of how many draws it took to cover all ten spaces and then for the next game they try to beat that total.

The importance of the conversations should be illustrated and reinforced by having students play several times in a large group setting before they try it with a partner.

Basic Facts
Lucky Guess and *Lucky Guesses*

Setting: All Students.

Materials: For **Lucky Guess**, a game sheet:
A piece of paper with the following equation for each of several rounds:
Round 1 $\square + \square = \square$.
Ten cards or pieces of paper in an envelope. Each card or piece
showing one of the numerals **0** to **9**.
Stickers with happy faces. ☺

To play *Lucky Guess* each player gets a game sheet with blank equations for several **Rounds**.

The players are told that the number names for zero to nine are on the ten cards in the envelope.

To ensure that the players know the number names that are in the envelope, the following requests can be repeated several times,

Name a number that is on the cards in the envelope.
Name a number that is not on the cards.

The goal of the game is to guess the names of two numbers that were drawn from the envelope and to use these number names to print a true equation in the blanks for each round.

Procedure
For each round two cards are drawn from the envelope, i.e., **3** and **8**.
The sum is calculated and announced to the students. The answer is **11**.

The players make one guess about what number names they think were drawn that would equal the announced sum, or **11**.

The players record their guesses and the answer in the boxes of the equation.

The statement **The answer is ___**, with the answer shown, in this case **11**, is recorded on a piece of paper or on the chalkboard.

The players are asked to suggest all of the true addition equations that they can think of for the answer, in this case **11**.

As the players suggest equations, they are recorded under **The answer is _11_**.

(cont'd next page ...)

Basic Facts
Lucky Guess and *Lucky Guesses* (continued)

This recording is repeated until there are no more responses to,
> *Who can think of an equation that is true which looks in any way different from those that have been recorded?*

5 + 6 = ☐ and **6 + 5 = ☐** are viewed as being different.

The two cards that were drawn are shown, the **3** and the **8**.

The players lucky enough to use the two number names **3** and **8** in either order receive or draw three happy faces.

The players who recorded different number names in their true equations receive or draw two happy faces.

Any player who recorded an equation that was 'close to being true' receives or draws one happy face, but the player is asked to explain how to change the equation to make it true.

Prior to each new round of a draw and making guesses, different players take turns trying to explain two ways, other than counting, that show that the answer is correct.

After three or more rounds of the game the happy faces are counted to find out who is *happiest* or *luckiest* and who is almost as *happy* or *lucky*.

Possible Variations
A second version of the game sheet, or *Lucky Guesses,* asks players to record two different guesses for each announced sum. It doubles the chances of guessing the correct addends and the students review two basic addition facts for each sum. The procedure is the same as for **Lucky Guess**.

The game sheet for each round shows: **Round 1**   ☐ + ☐ = ☐
                                                              ☐ + ☐ = ☐

The scoring for **Lucky Guesses** could be changed to:
- Four happy faces for recording the two number names as part of the first equation.
- Three happy faces for recording the two number names as part of the second equation.
- Two happy faces if different pairs of number names were used to record true equations.
- One happy face if a player is able to explain how an equation that is not true can be changed into one that is true.

Rather than drawing the cards at random two cards can be selected that are, for one reason or another, deemed appropriate for the players at that time.

To make doubles like **6 + 6 = ☐** part of the game setting, a deck of cards with two of each of the numerals **1** to **9** can be used.

Rather than using happy faces, points can be awarded.

Basic Facts
Lucky Cover-Up

Settings: Two teams or two players.

Materials: Two cardboard strips each marked off into ten spaces showing the
numerals from **1** to **10**, respectively.
The size of the spaces allows for the placement of chips, bottle caps,
blocks, pennies or buttons.
Two number cubes with **1** to **6** on the faces or two envelopes each
with six chips or pieces of cardboard showing one of **1** to **6**.
Chips, buttons, blocks, pennies or bottle caps.

The goal of the game is to cover more numerals than the opponent and that
the sum of the numerals left uncovered is less than the opponent's sum of uncovered
numerals.

Procedure
The number cubes are rolled or one chip is drawn from each of the envelopes.

For example, assume a **6** and a **5** are showing or are drawn. The answer is calculated,
6 + 5 = 11, and the answer is covered on the game board in whatever way possible and
in whatever way deemed most advantageous in terms of the goal of the game.

Some possibilities for covering **11** include:
   10 and **1**   **9** and **2**   **8** and **3**   **5, 4** and **2**   **6, 4** and **1**

Looking at the game board and examining it for possible combinations for a given sum
presents several opportunities for the strategy of *guess and test*. As this strategy is
employed, true equations as well as equations that are not true will be processed.

Ideally, the goal is to cover all numerals, but since that may not happen very often, the
sum of the numerals left uncovered is a player's score for the game.

When two students play against each other, they can take turns rolling the number cubes
or drawing one chip from each of the envelopes and watch each other making the moves
or, one student proceeds and goes as far as possible before the other student gets a turn.

Possible Variation
While playing the game, students are reminded to talk about all of the possible
combinations they try to come up with for a sum and explain why certain number names
are covered while others are not.

Older students can use the game to try and discover and describe strategies for the
moves they make and the strategies they think are favourable as far as the goal of the
game is concerned.

The game can be played without an opponent. A student playing this game alone sets out
to try to cover all number names, or to get a sum that is less than the one for the previous
try or tries.

Number Sense
Lucky Greatest Number

Setting: All students.

Materials: An envelope with ten disks or pieces of cardboard,
  each piece showing one of **0** to **9**.
  Each player has a game sheet with several of the following
  markings for each round of the game:
  Two attached boxes labelled **T** for tens and **O** for ones,
  one box for a discard for each round of the game.

Round 1
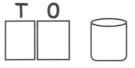

The goal of the game is to try and be lucky enough to end up with a name for
the greatest possible number or greater than the number name recorded by the other
players.

Procedure

For **Round 1**, a number name is drawn from the envelope. The name is announced.
Each player records the name in one of the three boxes. The name is returned to the
envelope. This procedure is repeated two more times with one opportunity to discard
a number name judged not to be in the best interest as far as the goal of the game is
concerned.

The students are asked to report their number names. These are recorded on the
chalkboard in order from greatest to least. Points are awarded. For example, five points
for the greatest, four for the next greatest, and so on. Students record their points
beside **Round 1** and the game continues with **Round 2**. After five rounds, players
calculate the total for the number of points they were lucky enough to obtain.

Possible Variations
  • The purpose of the game is changed to **Lucky Least Number**.
  • The number names drawn are not returned to the envelope.
  • Another place value position is added.
  • Another discard box is added.

Number Sense
Lucky Target

Setting: All students.

Materials: As for **Lucky Greatest Number**.

The goal of the game is to try and get as close to a target as possible.

Procedure
Two students are given the opportunity to create the target by entering their favourite single digit number name into one of the boxes. The students' names are printed below the target.

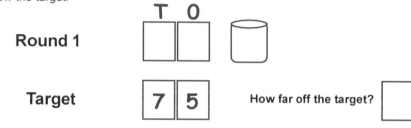

After a target has been identified, for example **75**, the following types of requests can be made before the round begins,
  - *Name a number that is very close to 75.*
  - *Name a different number name that is just as close as the last answer.*
  - *Name a number that is less than 75, and is far away from it.*
  - *Name a number greater than 75 but less than 100.*
  - *Name several number pairs that are the same distance from the target.*

The playing procedure is the same as for **Lucky Greatest Number**. After the recording is complete, students have to calculate how far off the target they are. All results of these calculations are recorded on the chalkboard in order from least to greatest.

Points are awarded, five for being closest to the target, four for being next closest, and so on. After several rounds the points are added.

After a numeral is drawn, several students could be asked to state where it was recorded and to share their reasons for recording it in that position.

Possible Variations
  - The goal of the game is changed to trying to stay as far away as possible from the target.
  - The target is thought of as a distance in centimetres or as an amount of money.
  - The target and the game sheet are extended to include hundreds for each round.

Number Sense and Addition
Lucky Greatest Sum

Setting: All students.

Materials: An envelope with ten disks or pieces of cardboard,
        each piece showing one of **0** to **9**.
    A game sheet for each player with the following markings
        for each round:

Round 1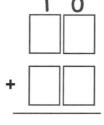

The goal of the game is to find out who is lucky enough to write the greatest sum.[5]

Procedure

Four numerals are drawn from the envelope. After each draw the numeral is announced and returned to the envelope. Each player records the announced numeral into one of the four boxes.

Each player calculates the sum. The different sums are recorded in order from greatest to least on the chalkboard. Points are awarded, i.e., five for the greatest sum, four for the next greatest, and so on. After several rounds, the points are added to determine the luckiest player.

Players in pairs are asked to find their answers to the following problems about the game:
- *What is the greatest possible sum?*
- *What is the next greatest possible sum?*
- *What is the least possible sum?*
- *What is the second least possible sum?*
- *What is the greatest and second greatest possible sum if the numerals are not returned to the envelope?*
- *What is the least and second least possible sum if the numerals are not returned to the envelope?*

Possible Variations
- The numerals are not returned to the envelope.
- The hundreds place value position is added to one or both of the addends.
- The goal of the game is changed to *Lucky Least Sum*.
- The goals of the game are changed to *Lucky Greatest Distance* or *Lucky Least Distance* or *Lucky Greatest Amount of Money* or *Lucky Least Amount of Money*.

Number Sense and Subtraction
Lucky Greatest Difference

Setting: All students.

Materials: An envelope with ten disks or pieces of cardboard,
each piece showing one of **1** to **9**.
A game sheet for each player with the following markings
for each round:

Round 1

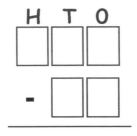

The goal of the game is to record the greatest possible difference
or a difference greater than the differences recorded by other players.

Procedure

Before the game is played, players in pairs can try to find the answers to
the following problems about the game,

- *What are the greatest and least possible differences if the
numerals are returned to the envelope?*
- *What are the greatest and least possible differences if the
numerals are not returned to the envelope?*
- *Explain your thinking about the top number and the number that
is subtracted in order to get the greatest difference. How would
your thinking change to get the least difference?*

The game playing procedure is the same as for **Lucky Greatest Sum**.
For this game five numerals are drawn.

Possible Variations

Some of the variations suggested for **Lucky Greatest Sum** can be adopted for this
game. One **Variation** can have every player record a **0** in the tens place of the
minuend and four numerals are drawn. This will ensure that renaming is required
and the players have to use flexible thinking about numbers before they calculate the
difference.

Basic Facts and Strategy
Target 22

Setting: All students or a small group or two players.

Materials: Two number cubes with **1** to **6** on the faces or two envelopes each
with six chips or pieces of cardboard showing one of **1** to **6**.
A game sheet for each player showing the following for each round:

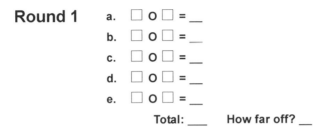

Round 1   a. ☐ o ☐ = __
b. ☐ o ☐ = __
c. ☐ o ☐ = __
d. ☐ o ☐ = __
e. ☐ o ☐ = __

Total: ___   How far off? __

The goal of the game is to try and get a total sum as close as
possible to **22**.

Procedure
Two number cubes are rolled or one chip is drawn from each of the two
envelopes five times. The numerals are announced and the players have one
of two choices, to write a true addition equation or write a true subtraction
equation.

After five turns, the answers for the five equations are added to find out how
close that total is to **22**.

Possible scoring procedures can include:
- The differences between **22** and the totals are the scores for
each round of the game. After several rounds these scores are
added to find out whose score is lowest or who managed to
come closest to the target.
- Points are awarded, the highest point for being closest to the
target. After several rounds the points are added to find out who
had the highest overall score and came closest to the target.

Students can be given the opportunity to explain their thinking as the game
progresses.

Estimating
Peeking at Paths

Setting: All students.

Materials: Paths consisting of two or three parts are drawn on pieces of cardboard.

The goal of the game is to estimate the lengths of paths.

Procedure

The students are assigned to five groups. For each round of the game a path is shown to each member of every group along with the unit that is to be used, i.e., paperclip; centimetre cube. The students are asked to estimate the length of the paths.

Charts on the chalkboard can be used for record keeping. One chart can show the estimates each group has decided on for each round. Another chart can show the points awarded to each group.[6]

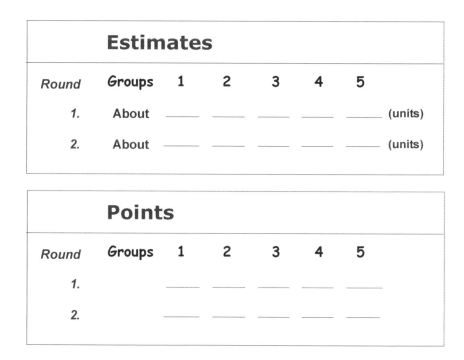

Estimates					
Round Groups	1	2	3	4	5
1. About	——	——	——	——	—— (units)
2. About	——	——	——	——	—— (units)

Points					
Round Groups	1	2	3	4	5
1.	——	——	——	——	——
2.	——	——	——	——	——

Every member of the group is requested to record an estimate on a piece of paper. Before one member of each group is called upon to record an estimate for the group in the chart on the chalkboard, the members are asked to try and reach consensus about what that recorded estimate should be.

(cont'd next page ...)

Estimating
Peeking at Paths (continued)

Younger students will require some guidance about how the decision on what to record might be made. Possible options to consider might be using the estimates made by two or more members of a group or reporting an estimate somewhere half way between the highest and lowest estimate.

The length of the path is measured to the nearest unit and the differences between the estimates and the measured length can be used as the score for that group. Rather than using the differences between the actual length and the estimates as scores, a **10 – 9 – 8 – 7 – 6** point system could be used where the highest score is assigned to the closest estimate.

After each round a few players can be given an opportunity to share the *estimation strategy* that they used. As a result of such discussions students are given the opportunity to employ different *estimation strategies* during future rounds. The game setting contributes to students becoming familiar with the centimetre as a unit of measurement of length.

Strategy
Take a Few - Not the Last One

Setting: Two players.

Materials: Two groups or rows of three counters.

The goal of the game is to get the opponent to take the last counter.

Procedure

The simple variation of the game NIM has two players facing the two rows of three counters. The goal of the game is to be the winner by having the other player end up with the last counter.[7]

Turns are taken. For each turn a player removes counters. The only proviso is that a player must remove fewer counters than there are in a row at the beginning of the game. That means that for this game setting the options of removing counters are one or two.

Possible Variations
- The number of counters in each row is increased to four.
- The number of rows is increased to three.

The players can be asked to prepare frequency tallies or a bar graph for:
- *Who won the game?*   **First Taker**   **Second Taker**
- *How many were taken at the start?*   **1 2 3 (row)**

While players are observed in action, they could be asked to explain their reasons for taking as many counters as they did.

Use of Imagination
Inventing Games and Game Settings

The request to have students make up their own game gives them the opportunity to be creative.[8] Such a request can include the following decision making aspects:

- **A title**: What do you think is a good title, one that might make other students curious about your game or that tells them what the game might be about?
- **A general goal**: What should those who will play the game with you remember about playing games?
- **Specific goals**: What should players keep in mind while playing your game with you?
- **Rules**: What are the rules that need to be followed?
- **Decorations**: Are there some ways to decorate your game that would show players what the game is about or what some of the rules might be?

When the rules for a game are tested for the first time while playing it with another student, the questions that are posed will likely result in revisions. Additional rules may be added. Rewriting and revising the rules involves further reflection.

Possible Materials

- **Activity sheets**: The example included in **Chapter 6 – Basic Addition Facts** (see p. 97) illustrates that the suggestion to use a practice sheet to make up a game can result in fascinating creations:
  - ▸ Students recorded general goals: *No cheating*; *Good luck*.
  - ▸ There were examples of specific goals: *The idea of the game is not to punch people – not to push people – not to scream at people.*
  - ▸ Students made up their own titles: *Pick It*; *Adventure*.
  - ▸ Rules were created, tested and then edited.
  - ▸ Some students decorated their game boards and provided additional information about their games.[9]

- **Piece of cardboard and felt pens**: These materials will require a few initial suggestions. For example, students could be asked to make a sketch that shows a lake, a house and several paths with rocks that are on the paths. A reminder about making up a game that makes use of something from the mathematics classroom, i.e., numbers, number names, blocks, 2-D figures, or fractions, may also be in order since some students will end up inventing a fairy tale setting that does not include any mathematics.

- **Area carpet**: Some classrooms have area rugs with colourful designs or pictures. Pairs of students could work together to make up rules for a game that can be played on the carpet. The game is explained to classmates who are requested to think of questions they have about the rules. The authors of the game use these questions to revise their rules.

(cont'd next page ...)

Use of Imagination
Inventing Games and Game Settings (continued)

- **Familiar game boards**: The students are asked to think of a game they have played and like to play. If they were to design a different game or make up a few new rules for the game, what would they be?

- **Unfamiliar games and boards**: The students are asked to look at a game board and the pieces that come with it, i.e., Chess, Backgammon, and invent their own game for the setting. What do they think the rules should be for their games on these boards?

The orchestration of these settings can result in having students *think* as well as *think about their thinking*. Appropriate questions and follow-up tasks can *advance* students' *thinking*. Many opportunities exist to foster the development of *communication* skills.

Assessment Suggestions

Since there are no specific goals related to playing and inventing games the assessment information that can be collected and shared with parents is of a general nature. The indicators that can become apparent during the activities related to games can provide valuable information about several desirable characteristics that are related to success with mathematics learning. These include indicators of:

- *willingness to talk.*

- *willingness to try different strategies.*

- *thinking* and *thinking about thinking.*

- *flexible thinking.*

- *curiosity.*

- *spontaneity.*

- *connecting.*

- *use of imagination.*

- *high self-esteem.*

Reporting

The observations that are made as students gather data, organize data, draw conclusions and share results with their classmates and as they play and invent games provide information about the critical components that students must encounter in a mathematics program.

Parents can be informed about possible indicators of

- *communicating* orally as well as in written form – interpreting and sharing data.
- *mathematical reasoning* – drawing conclusions; making up meaningful rules.
- *visualizing* and use of *imagination* – possible interpretation of a graph with missing information; designing a game for a game board.
- *confidence*, *risk taking* and *perseverance* – making up different types of questions for a display; sharing invented rules and revising the rules.

For Reflection

During a meeting with parents you plan to make the suggestion that they use some data about their children and draw a graph, i.e., charting their growth; charting the growth of a plant or a pet. What sample tasks and sample questions for parents would you include on a handout to them? What are the main reasons for these suggestions?

What are the main points you would include in a presentation to parents that is entitled: *Playing Games with Your Children*?

Selected References

Chapter 1

(1) Reys, R. (1971). Consideration for teachers using manipulative materials. *Arithmetic Teacher*, 18(8), 551-558.

(2) Ministries of Education. (2006). *The Common Curriculum Framework for K to 9 Mathematics – Western and Northern Canadian Protocol.*

(3) Willoughby, S. (1990). *Mathematics for a Changing World.* Alexandria, Virginia: Association for Supervision and Curriculum Development.

(4) Polyani, M. (1989). Math Studies Should Teach Problem Solving. *Education Leader*, (June), 1-16.

Chapter 2

(1) Charles, R. and Lobato, J. (1996). *Future Basics: Developing Numerical Power.* Monograph. Golden, Colo.: National Council of Supervisors of Mathematics.

(2) Hiebert, J. (2000). What Can We Expect from Research? *Teaching Children Mathematics*, 6(7), 436-437.

(3) Davis, R. (1986). *Learning Mathematics – The Cognitive Science Approach to Mathematics Education.* New Jersey: Ablex Publishing Corporation.

(4) Greenwood, J. (1993). On the Nature of Teaching and Assessing "Mathematical Power" and "Mathematical Thinking." *Arithmetic Teacher*, 41(3), 144-152.

(5) Flewelling, G. and Higginson, W. (2000). A handbook on rich learning tasks (Realizing a vision of tomorrow's mathematics classroom). Faculty of Education, Queens University: Kingston, Ontario, Canada.

(6) Bruni, J. and Seidenstein, R. (1991). Geometric Concepts and Spatial Sense. In J. Payne (Ed.) Mathematics for the young child. Reston, VA: The Council.

(7) Reys, R., Suydam, M. and Lindquist, M. (1989). Helping Children Learn Mathematics. Toronto: Prentice-Hall Canada, Inc.

Chapter 3

(1) Suydam, M. and Weaver, J. (1977). Research on Problem Solving: Implications for Elementary Classrooms. *Arithmetic Teacher*, 25(2), 40-42.

(2) Suydam, M. (1980). Untangling Clues from Research on Problem Solving. In S. Krulik (Ed.) *Problem Solving in School Mathematics.* 1980 Yearbook of the National Council of Teachers of Mathematics, 34-50. Reston, VA: The Council.

(3) Wirtz, R. and Kahn, E. (1983). Another Look at Applications in Elementary School Mathematics. *Arithmetic Teacher,* 31(2), 15-21.

(4) Willoughby, S. (1990). *Mathematics for a Changing World.* Alexandria, Virginia: Association for Supervision and Curriculum Development.

Chapter 4

(1) Spungin, R. (1996). First and second grade students communicate mathematics. *Teaching Children Mathematics*, 3(4), 174-179.

(2) Liedtke, W., Kallio, P. & O'Brien, M. (1998). Confidence and risk taking in the mathematics classroom (Grade1). *Primary Leadership*. 1(2), 64-66.

Chapter 5

(1) Ministry of Education. (2000). *The Primary Program – A Framework for Teaching*. Victoria, B.C.: Evaluation Branch, Ministry of Education.

(2) Charles, R. and Lobato, J. (1996). *Future Basics: Developing Numerical Power*. Monograph. Golden, Colo.: National Council of Supervisors of Mathematics.

(3) Liedtke, W. (1983). Young Children – Small Numbers: Making Numbers Come Alive. *Arithmetic Teacher*. 31(1), 34-36.

(4) Liedtke, W. (2004). A Focus on Number Sense Makes a Lot of Sense. *delta-K*. 41(2), 15-17.

(5) Gupta, A. (2009). Math is never wasted on the young. *Vancouver Sun* (The Daily Special). March 25, A9.

(6) Williams, T. (2009). Letter. *Vancouver Sun* (The Daily Special). April 1, A9.

(7) Mighton, J. (received 2008). INTRODUCTION TOO THE JUMP WORKBOOKS. Handout, 1-14

Chapter 6

(1) Charles, R. and Lobato, J. (1996). *Future Basics: Developing Numerical Power*. Monograph. Golden, Colo.: National Council of Supervisors of Mathematics.

(2) Campbell, P. (1981). What Children See in Mathematics Textbook Pictures? *Arithmetic Teacher*, 28(5), 12-16.

(3) Liedtke, W. (1998). *Interview and Intervention Strategies for Mathematics*. Sherwood Park, Alberta: ECSI Publishing Inc.

(4) Liedtke, W. (2004). Why Do Numerate Students and Adults Lack Conceptual Understanding of Division? *Vector*, 45(1), 21-30.

(5) Cathcart, G., Pothier, Y. and Vance, J. (2004). *Learning Mathematics in Elementary and Middle Schools* (4th edition). Toronto, Ontario: Pearson Education Canada Inc.

(6) Howe, R. (1999). Knowing and teaching elementary mathematics (Book Review). *Journal for Research in Mathematics Education*, 30(5), 579-589.

(7) Liedtke, W. (1999). Riddles and Games – Selected Thoughts and Observations. *Canadian Children*, 24(1), 20-24.

(8) Liedtke, W. and Sales, J. (2008). Writing Tasks That Succeed. In P. Elliot and C. Elliot Garnett (Eds.) *Getting into the Mathematics Conversation – Valuing Communication in Mathematics Classrooms*. Readings from NCTM's School-Based Journals, 334-341.Reston, VA: The Council.

Chapter 7

(1) Wheatley, G. (1990). Spatial Sense and Mathematics Learning. *Arithmetic Teacher*, 37(6), 53-58.

(2) Bruni, J. and Seidenstein, R. (1991). Geometry Concepts and Spatial Sense. In J. Payne (Ed.) *Mathematics for the Young Child,* 202-227. Reston, VA.: The Council

(3) Special Issue (1990). Spatial Sense. *The Arithmetic Teacher*, 37(6).

(4) Liedtke, W. (1993-94). Teaching Geometry – Developing Spatial Sense. *Prime Areas*, 36(3), 56-64.

Chapter 8

(1) Liedtke, W. Measurement. (1990). In J. Payne (ED.) *Mathematics for the Young Child*, 228-249. Reston, VA.: The Council.

(2) Shaw, J. & Puckett Cliatt, M. (1989). Developing Measurement Sense. In P. Trafton (Ed.) *New Directions for Elementary School Mathematics*, 149-155. Reston, VA.: The Council

(3) Liedtke, W. (1993-94). Teaching Measurement: Promoting the Development of Measurement Sense. *Prime Areas*, 36(1), 98-103.

(4) Copeland, R. (1979). *How Children Learn Mathematics*. New York: Macmillan Publishing Co.

Chapter 9

(1) Liedtke, W. (1992). Versatile Graphing. *Vector*, 33(2), 34-39.

(2) Liedtke, W. & Vance, J. (1978). Simulating Problem Solving and Classroom Settings. *Arithmetic Teacher*, 25(8), 35-38.

(3) Liedtke, W. (1980). Games for the Primary Grades. *The Arithmetic Teacher*, 28(4), 30-31.

(4) Willoughby, S. (1990). *Mathematics for a Changing World*. Alexandria, Virginia: Association for Supervision and Curriculum Development.

(5) Liedtke, W. (1998). *Interview and Intervention Strategies for Mathematics*. Sherwood Park, Alberta: ECSI Publishing Inc.

(6) Liedtke, W. Measurement. (1990). In J. Payne (ED.) *Mathematics for the Young Child*, 228-249. Reston, VA.: The Council.

(7) Liedtke W. & Nelson, D. (1974). *Mathematical Experiences – Primary Division*. Toronto, Canada: Encyclopaedia Britannica Ltd.

(8) Liedtke, W. (1999). Riddles and Games – Selected Thoughts and Observations. *Canadian Children*, 24(2), 20-24.

(9) Liedtke, W. (1996-97). Fostering the Development of Conceptual Knowledge: The Basic Addition Facts. *Prime Areas*, 39(1), 42-51.